URBAN PROCESS AND POWER

Urban Process and Power not only explains the processes fashioning the built environment but also shows how they reflect the dominant social and political values and the economic circumstances in which they are taking place.

Policies towards the built environment during the 'neo-liberal' period since the mid-1970s have increased inequalities, wasted resources and have frequently been self-defeating even in terms of neo-liberalism's own values and strategies. The author argues that these effects are serious and that they raise questions of accountability since they are taking place in the context of an accelerating de-democratisation of the urban development process.

Comparison with processes elsewhere in Europe helps to confirm the view that recent deregulatory strategies have produced many adverse effects and underlines the need to move back to land development procedures that show a better balance between the private, public and voluntary sectors and a greater sensitivity to users of the environment.

The book will be invaluable to those interested in the development of the built environment and especially to those concerned with the extent to which democratic accountability is being eroded.

Peter Ambrose is Senior Lecturer in Social Policy and Director of the Centre for Urban and Regional Research at the University of Sussex.

URBAN PROCESS AND POWER

Peter Ambrose

London and New York

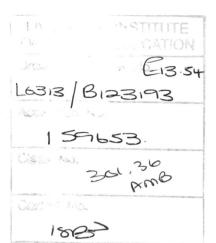
First published 1994
by Routledge
11 New Fetter Lane, London EC4P 4EE

Simultaneously published in the USA and Canada
by Routledge
29 West 35th Street, New York, NY 10001

Typeset in Garamond by Solidus (Bristol) Limited

Printed and bound in Great Britain by
Biddles Ltd, Guildford and King's Lynn

British Library Cataloguing in Publication Data
A catalogue record for this book is available from the British Library

Library of Congress Cataloging in Publication Data
Ambrose, Peter J.
Urban process and power/Peter Ambrose.
p.   cm.
Includes bibliographical references and index.
1. City planning. 2. Urban policy. I. Title.
HT165.5.A43   1994
307.76–dc20   93-44565

ISBN 0-415-00850-6 (hbk)
ISBN 0-415-00851-4 (pbk)

To
GUY, STEPH AND DAVID
with love,

and to poor, battered
COMPASSION –
may it soon make a come-back
in public affairs.

CONTENTS

ILLUSTRATIONS

FIGURES

PLATES

TABLES

ACKNOWLEDGMENTS

The following, listed in alphabetical order, have all been helpful as friends, colleagues or sources of information and ideas: Barbara Ainger, Jenny Backwell, James Barlow, Ken Bartlett, Edward Burd, David Cadman, John Cook, Peter Dickens, Simon Duncan, Chris Ellmers, Stephen Hill, Aldous Hodgkinson, Jerry Le Sueur, Alan Murie and Tony Trott. The author is also grateful to Keith Hunt for taking some of the photographs, to the staff of Topham Picture Source for helping in the search for others and to Susan Rowland for the impeccably drawn diagrams. At Routledge, Tristan Palmer, Sue Bilton and Sarah Lloyd have been very supportive as editors. Andy Thornley made full and most helpful comments on an earlier draft of the book. My wife Sue was a source of much love and support at a time when she was working for her own finals – if the book is anywhere near the quality of her excellent First I will be happy. All these people have helped make the book better than it would have been. Its shortcomings, of course, stem from the author alone.

INTRODUCTORY NOTE

This book grew out of the author's experience of teaching the Urban Studies undergraduate degree at the University of Sussex. He and Peter Dickens, with genial and expert help from Fred Gray, Pete Saunders, Mick Dunford, Ray Robinson, James Barlow, Simon Duncan and other colleagues, kept a small show on the road for nearly twenty years until pressures for rationalisation finally killed it off in the early 1990s – not before it had graduated a lot of people who are now doing a lot of useful work. In a sense this book stands as a memorial to the degree – probably an unworthy one.

As the Thatcher years dragged on, the author, together with many others, was becoming very worried about the way in which unelected bodies were steadily usurping powers over the evolution of the built environment, powers that had previously been exercised democratically. As a London Docklands watcher since 1972, and a Millwall supporter for long before that, he was especially concerned with the activities of the London Docklands Development Corporation. But that was only the tip of the iceberg. The progressive loss of local authority power to centralist dictates and non-elected bodies of various kinds was all part of a picture.

As the book moved towards its conclusion in the spring of 1993 it became obvious that this concern with the loss of democracy was becoming widespread. In particular Patrick Wintour wrote a piece in the *Guardian* (29 April 1993) about the 'Relentless advance of the "quangocracy"'. He detailed the growth of bodies that were turning Britain into an 'unelected state' or a 'new magistracy' based on local patronage. There were sixth-form college boards spiced with Tory parliamentarians. The control of hospitals was being removed from health authority boards on which sat local councillors, to trusts with centrally appointed members. Local councils were losing the right to nominate members to the governing bodies of further-education colleges. Local educational authorities stood to lose perhaps half the country's secondary schools to central grant-maintained status. The training role of the Manpower Services Commission was being replaced by Training and Enterprise Councils Ltd. Local authorities were losing control of police committees. Local authority housing departments were losing their capability to provide directly for local housing needs and were obliged to work increasingly through non-elected agencies. It was reported in the article that some ministers, even Graham Mather of Tory 'think tank' fame, felt that the pendulum might have swung too far.

What is happening here? It seems clear that the system of local democracy, set up just about a century ago, is being dismantled. We ignore these tendencies at our peril. Consumers' 'charters', advanced as the panacea by apologists for what is happening, do not make good what is being lost via the ballot box. The real Chartists, some of whom gave their lives a century and a half ago to help win our democratic rights, would not be taken in by today's toothless charters; nor would those who fought for votes for women early this century.

This book seeks to highlight the de-democratisation process in one particular field, the shaping of the built environment, and to lend such weight as it has to swinging the pendulum the other way.

A GUIDE TO THE BOOK

This book has two main aims. One is to analyse and explain, in as value-free a manner as possible, the processes that produce and continually re-produce the urban environment in which most of us live. The second and more political aim is to focus especially on recent changes in the *control* of these processes and on the ideological belief system that has brought these changes about. The book takes as one starting point the immense disparities that are shown to exist between two of the 'worst' and one of the 'best' urban areas in Britain. It seeks later to illuminate how and why these differences arise and how and why they are perpetuated. It also argues that there is a relationship, intuitively plausible if not yet demonstrable, between the growth of these disparities and the decreasing extent to which environment-shaping processes are subject to democratic control.

The adjective 'urban' is not intended to have any precise meaning in this context. It should not be taken to imply that the book is exclusively about settlements of any given administrative status or above any particular population size, since most of the processes discussed occur also in villages or rural situations. In any case, in complex modern societies such as Britain the terms 'urban' and 'rural' have lost much of their social or analytical meaning so far as the lifestyle of the vast majority of the population is concerned. The density of human occupancy may continue to vary from very dense to very sparse but the variations in experience and culture that Louis Wirth (Wirth, 1938) was able to differentiate so clearly a half-century ago no longer exist to anything like the same extent. It would be more accurate to use the term 'built environment' to denote that the analysis applies to any area where development has taken place on a significant scale. But this would have precluded the use of what is hopefully a snappy title and a neatly corresponding chapter scheme. Perhaps some poetic licence can be allowed even in dreary textbooks.

Part I of the book seeks in Chapter 1 to identify some key characteristics of the urban environment and to show especially that it is the continually changing product not of 'natural forces' but of consciously directed human action. In Chapter 2 some statistical material is presented to demonstrate the extent of the differences in the quality of the urban environment between three selected local authority areas – Knowsley (outer Liverpool), Tower Hamlets (inner London) and Wokingham (Berkshire). How do these differences arise, and how are they perpetuated? This chapter may be skipped, or used simply as a source of data, by those already well aware of the wide disparities in urban conditions in post-Thatcher England.

Part II analyses the three main production processes that generate these varied environments. Chapter 3 presents a general model outlining the stages in the life of any development and the agencies that play a part at each stage. Chapters 4 and 5 discuss the two main forms of development carried out by profit-seeking promoters – those producing buildings to rent, and those producing buildings for sale. Chapter 6 then deals with publicly promoted development undertaken for 'social' purposes, and Chapter 7 with some other 'non-profit' forms of development. Between them these four chapters are intended to cover virtually all development processes in a reasonably objective manner. It so happens that the current output of the processes covered in Chapter 4 is largely shops, offices and industrial premises, while the processes discussed in Chapters 5, 6 and 7 currently produce primarily housing. This has not always been the case, and it is an important part of the book's logic that the processes producing new built environment are best analysed by mode of development rather than by form of output.

Part III then introduces the 'political variable' into this hitherto largely technical analysis. Chapter 8 examines the ideological and political framework provided by the ascendancy since the early 1970s of the 'New Right' and it offers a critique of some key neo-liberal arguments. Chapter 9 traces the effects of the policies carried out by the neo-liberal administrations of the 1980s on the output of new built environment, on rent-setting principles and policies, and on specific development issues such as London's Docklands. It returns to the point that there was both an increase in inequality and a significant process of de-democratisation during the Thatcher years. This lends some weight to the proposition that there is some causal connection between the two.

Part IV discusses development processes elsewhere and seeks to identify lessons that may be learned. Chapter 10 presents some comparative material and considers the balance of 'state', 'market' and 'citizen' in the land-development systems in several other West European nations. The criterion for inclusion of the examples is about as unsystematic as it could be – these were simply situations of which the author had some first-hand knowledge. The chapter considers practices in both Sweden and Denmark which aim to empower residents in housing development and management issues. It also gives a brief account of the activities of the British Housing Advisory Programme in Bulgaria. These focus partly on the encouragement of more citizen participation in development and housing management processes in the new post-socialist situation. Finally, Chapter 11 seeks to draw some general conclusions from the material presented in earlier chapters. For example, is there a trade-off between 'economically optimal' and 'equitable' approaches to the solution of urban problems? In view of all the complexities, to what extent is it reasonable to expect that the processes shaping the built environment *should* involve the participation of users? And if they should, what 'model' might best serve future needs – formal electoral procedures or those plus more fully developed 'pluralist' and 'voluntary sector' activity?

Some of the chapters, especially in Part II, spend considerable time on the evolution of development processes over the period, roughly, of the past century. This is because the built environment is a cultural product which has grown gradually by a process of accretion. It cannot fully be understood except as a reflection of the social, political and economic circumstances in the period over which it was formed. This particular timespan has been chosen because most of the built environment we experience has been built during this period. This discussion of general processes is complemented by case-study material – 'corroborative detail intended to give artistic verisimilitude to an otherwise bald and unconvincing narrative', as Pooh-Bah has it in *The Mikado*. Some of this detail is found in the text, some in the case-study material and some in the lengthy captions to the illustrations.

Part I

URBAN
Why and how do urban areas vary?

The aim of the first part of the book, which contains two chapters, is to draw attention to a number of key characteristics of the urban environment as a product of consciously directed human activity and to demonstrate the extreme differences in the quality of three selected urban environments.

Chapter 1 discusses the nature of the built environment as a human product in a continual state of change. It considers the processes that produce inappropriate local 'mixes' of development – especially local housing shortages. It argues that adequate and accessible housing should be seen as a central element in the infrastructure of an area – equally important with, for example, power supply and roads as a precondition for optimum regional economic performance. The chapter considers the two key concepts 'democracy' and 'free market' and raises the issue of whether they are synonymous or contradictory ideas. Finally, given the recent changes in the balance of influence between 'state' and 'market' in shaping the urban environment, it questions whether and on what grounds we should argue for a democratic input into the processes that produce urban development and redevelopment.

Chapter 2 seeks to identify the current degree of inequality in the quality of local urban environments in England. It presents a set of statistical data drawn from standard sources to compare three sharply contrasting local authority areas, Tower Hamlets (a London borough), Knowsley (a metropolitan borough in outer Liverpool) and Wokingham (a district of Berkshire). The comparison is arranged under four headings: demographic structure and change, housing conditions and costs, employment/occupational structures and unemployment, and finally standards of educational provision and attainment.

This first part of the book seeks to show that the urban environment is a tangible artifact which mirrors the power relations in society. It quantifies some of the more important disparities in the quality of the urban environment and provides a context for a number of key issues which are addressed in the remainder of the book. How much inequality is too much? What processes generate such disparate environments? How are the processes affected by changes in dominant ideologies and politics – especially those in the period since the late 1970s? Who controls the processes? Are the control mechanisms 'democratic'? Can they be and should they be? Are they becoming less or more so? Do arrangements elsewhere in Europe confer greater degrees of participative democracy in the processes that produce and manage the built environment? Can we learn anything from them? What advice can we usefully pass on to the 'reforming' countries of East and Central Europe?

1

'HUMAN NATURE' AND THE URBAN ENVIRONMENT

The city is not ... merely a physical mechanism and an artificial construction. It is involved in the vital processes of the people who compose it; it is a product of nature, and particularly of human nature.

(Park, Burgess and McKenzie, 1925, chapter 1)

There are a number of initial points to make about the built environment, some of them summed up succinctly in this quotation from three of the pioneers of urban analysis. First its 'mix' and distributional pattern are all-pervasively important to the quality of our lives – collectively and individually. These characteristics are fashioned and constantly transformed by purposeful human action and, in turn, they fashion and transform people. The differences in the quality of urban environments are shown in Chapter 2. The degree of segregation of different qualities and tenure-forms of housing and the extent to which 'up-market' and 'down-market' areas are spatially differentiated contribute to the imaging of the social hierarchy and the self-imaging of one's place within it. Those who doubt this should read about the incredible saga of the wall built by owner occupiers in a suburb of Oxford in the 1930s to separate themselves off from the adjacent council estate (Collison, 1963).

The 'mix' of built space available in an area is often socially and economically inappropriate. Large amounts of retail and office space which cannot be let or sold often co-exist with an acute shortage of housing of acceptable quality. The over-supply of the one reflects at the least a temporary loss for investors and developers, while the shortfall of the other generates social hardship and heavy public cost in terms of emergency arrangements. Both effects waste resources on a scale that the British economy can ill afford, apart from causing considerable human misery. How does this mix get so badly out of balance?

The basic problem here is that the processes of capital investment and

Plate 1.1 Dramatic transformations of the urban environment. The scene would have been totally different even ten years previously. This view looking east along Rotherhithe Street in London's Docklands shows, from right to left, a small block of previously local authority flats now much refurbished, some inter-war terraced housing, the Canary Wharf and other commercial development schemes (across the river in the Isle of Dogs), Barratt's recently built 'Sovereign View' development of riverside houses and flats for sale, and the site being prepared for later phases of the same scheme. Previously in this vicinity there had been many wharves, including Pageant and Lavender Wharves handling the import of numerous goods, a London County Council Fire Station, and a Port of London Authority pumping station. Photograph by Keith Hunt.

disinvestment that add or subtract jobs in a local area often have effects more rapidly than the processes that produce social infrastructure in the form of the housing, schools and similar facilities necessary for the reproduction of the labour force. Characteristically, in a mixed economy such as Britain's, the changes in jobs stem largely from profit-seeking investment by one particular category of capital – manufacturing and service-sector entrepreneurs. The 'social infrastructure' evolves via an often poorly co-ordinated response by a different category of capital (private housebuilders) in combination with public sector activities undertaken in order to discharge statutory responsibilities. In recent years, roughly since the mid-1970s, the capacity of local public agencies to provide a reasonable share of this 'social infrastructure' has been materially diminished and it has been left more to the market to provide the housing to match the jobs. This has often led, as Chapter 9 will show, to local shortages of low-rent/price housing which, in turn, may well inhibit the economic development of the areas in question.

It seems fair to argue that the relationship between economic development

4

Plate 1.2 Another example of rapid transformation, looking east across the Eastern Branch of Poplar Dock. This was begun in 1852 for the North London Railway and closed in the early 1980s. The block of flats is Alberta House, built by the London County Council as workers' housing around 1939. The pub was built soon after. The railmounted electric quay crane dates from the 1950s. The large white structure is an air vent for the Blackwall Tunnel constructed in 1963–7. Behind is the Reuter Docklands Centre designed by the Richard Rogers Partnership and opened in 1989. It straddles what had been Blackwall Engineering Yard's dry dock. The Yard was sold in 1987 to benefit from the high land prices, thus ending over 370 years of continuous shipbuilding and repair on this site. Photograph by Keith Hunt.

and the quality of the built environment has been insufficiently understood and consistently undervalued. This may be because the size and apparent inevitability of the processes involved serves, paradoxically, to reduce their visibility. Few would dispute the vital significance of a proper level of investment in forms of infrastructure that ensure power supplies, water distribution, telecommunications or transport systems – all of which are clearly relevant to economic development and social welfare. But an adequate supply of good, comfortable and safe housing is an even more significant form of infrastructure since it is vital to health, comfort and the development of human capabilities (see Ineichen, 1993) – and therefore to the economic performance of the society. While investment in the supply of power, water, telephones and transport has risen in real terms, public capital investment in housing has fallen sharply since the mid-1970s. The output, of both private and public housing, has slumped drastically in the late 1980s and the 1990s, and housing conditions for many have materially worsened. Why is this? And who has decisive power to shape these patterns of investment?

There is another initial point to make. The built environment, despite its

5

solid appearance, is dynamic rather than static. It is in a constant state of change. The changes range from the barely perceptible and inevitable, the gradual deterioration of a building, to the very dramatic and contentious. A city-centre site can be redeveloped over a two-year period, a bypass can transform the landscape within perhaps five years, and whole inner city areas can be re-fashioned, physically and socially, over a ten-year timescale. In Chapter 9 we will consider one striking example of such speedy transformation and its effects on local residents. Governmental intervention in the form of 'planning' and 'conservation' may affect the nature and pace of these changes, and may even give them a degree of democratic accountability (how much is one of the key concerns of the book). But changes to the form of the built environment have always happened anyway. Nothing fashioned by people lasts forever.

The next key point is that the changes are purposeful. They do not happen without a considerable amount of conscious forethought on the part of the individuals, groups or interests who are promoting and carrying them out. This has been true all through history, not just since the era of formalised land-use 'planning' began (in the case of Britain in 1947). The reason is obvious. Refashioning a part of the built environment involves money, usually a lot of money. It is a characteristic feature of land development and redevelopment that this investment has to take place 'up front' and that the benefit flows over a subsequent timescale. The realisation of the benefits, the final stage of the process, takes a number of forms, most of which will be subsequently discussed. But the main point is that changes in the built environment are initiated only after much forethought and careful calculation about the costs involved and the benefits to be derived – whether these are expected to be in terms of financial accumulation, personal utility or possibly 'public benefit'.

WHO FASHIONS THE BUILT ENVIRONMENT?

So who precisely *are* the individuals, groups and interests who initiate and organise the changes to the built environment? Here the 'us' in the second sentence of the Guide to the Book, if left unconsidered, could de-politicise the analysis and thus negate one of the main aims. 'We' may all live in the built environment, but probably most of us feel that we have little control over the forces that shape it. In fact it is reasonable to say that the initiators and organisers of change, the active modifiers of the built environment, are not usually conceived of as 'us' at all but as 'they'. But it is usually some shadowy undifferentiated 'they'. 'They' are building a new retail/leisure complex or demolishing houses for a road widening scheme. Or if it's not 'them' it's 'the planners'. But who really are 'they'? It is a rich irony that the built forms in which we spend most of our waking hours, and which can add to or detract so much from our comfort, appear, so far as most people are

6

concerned, to emerge via some hazy process carried out by some anonymous 'they'.

There are a number of reasons for this. Often changes in the environment are projected by the media or popular literature as being the result of some 'natural' process or the inevitable consequence of the adoption of some new technology rather than as the product of a set of carefully thought-out investment decisions. The issue is de-personalised; inanimate objects are curiously endowed with animate characteristics. It is 'the city' that is expanding outwards into agricultural land. It was 'the railways' that came in the middle of the last century and took so much city-centre land for their termini. It is 'the motor car' that is generating the demand for out-of-town shopping centres. Even in the professional understanding of many policy-makers there is a predominantly passive mindset in the face of some very active, profit-driven processes. There are assumed to be 'city-centre' land uses – activities such as retailing and office-based employment – that 'naturally' tend to congregate in and near city centres. If pressed many would say that this phenomenon occurs because city-centre land values are high and these uses win out in a process of competitive bidding for sites.

But this is circular and technically incorrect. The processes are not natural; they are controlled by people. The uses that can occur in these, or any other, locations are those specified by the planning system. Decisions to guide the pattern of land uses are, or should be, made by elected members of planning committees advised by professionally trained planning officers. They are often, in turn, responding in some way or another to commercial develop-ment pressures. If it is all to do with 'natural forces' where does that leave their decision-making capability – and our votes in local elections? In any case, even the most cursory examination of a 'socialist' city, say East Berlin before 1989, reveals mixed land uses in the central areas, with a strong presence of residential use. In other words the pattern regarded as 'natural' in the West is no such thing – it is politically and financially fashioned by the interplay of largely profit-seeking forces in the land market.

This book, then, is an attempt to sharpen and politicise the understanding of the processes that produce the urban environment and especially to examine their relationship to the democratic process. The attention paid to these issues in the formal educational curriculum from primary to higher education is derisory. In junior school children seem to learn more about the nature and extent of the Roman Empire than they do about local housing provision, where jobs come from or who carries out new retail developments and why. Later on, at secondary level, an adequate study of the built environment and urban problems simply cannot be squeezed into the existing 'subjects' such as History, Geography and Sociology. Although things are changing, administrative inertia still works to organise school curricula by these 'subjects' rather than by the issues and problems that actually arise in the real world. To some extent one could make the same judgment about the higher-education sector, despite the welcome growth of interdisciplinary degree courses in recent decades. Someone, it was probably David Harvey, once made the point that disciplinary boundaries are inherently counter-revolutionary. This may be reassuring to some, since it is often said that most people are in favour of all revolutions except the next one. But disciplinary-based myopia is often an impediment to a critical understanding of the environment-shaping processes.

It is now especially timely to draw attention to these processes, and to raise the question of the recent decline in their democratic accountability. It has been evident over the modern era of urban development, roughly since the Industrial Revolution, that decisions concerning the initiation and removal of jobs, and those concerning the provision of housing and other built environment, have increasingly been taken at a level beyond the comprehension and control of the local employee and resident. Industrial capital now operates on a global scale, not the parish or town scale of the eighteenth century. Finance capital, which had very limited functions two hundred years ago, is now also organised on a world scale (see Budd and Whimster, 1992). Its original function was to finance and insure mercantile transactions and trade between nations. This has been supplemented by an investment function which, in a 'tail wagging the dog' manner, helps to fund new built environment such as Canary Wharf in London's Docklands – sometimes before the demand for the development has become apparent. The private housebuilding industry, though highly fragmented by modern standards, is still to some degree dominated by the 'volume' builders. The initiation and construction of motorways and other major roads is largely a central-government function. Where is the voice of the individual in all this, where the capability of the local council? How decisive even is the power of national governments in the face of international financial agencies who may insist on changes in domestic public investment strategies as a condition of support for the economy, as the International Monetary Fund and World Bank are perfectly capable of doing?

'DEMOCRACY' AND THE 'FREE MARKET' – SYNONYMS OR OPPOSITES?

At this particular juncture there is an even more pressing reason to examine and seek to redress this long-term process of citizen disempowerment. The formerly socialist members of the 'Eastern bloc' are reorganising their social and economic systems. Following over forty years of highly centralist, externally dependent, administration they are very keen to become 'normal' countries, to use the phrase often heard from their leaders. They wish to develop 'free market' economies and aspire to join the European Community. In the first few years after 1989, when the Austria/Hungary border opened up and the Berlin Wall was breached, there was a heady but understandable belief that the 'free market' would re-assert itself and that strategies such as the decontrol of prices and rents would solve all the problems. The 'transitional' countries wished to move simultaneously towards 'democracy' and the 'free market' – as if those two concepts were unproblematically synonymous. They all, especially the Poles, saw Mrs Thatcher as the embodiment of these two ideals – and of much else that was desirable besides.

But to an extent this vision is illusory and the proposed trajectory was not practicable in its initially conceived form. Much that happened in the Thatcher/Reagan years was not what it seemed to these observers. In Britain increasingly centralist interventionary policies have been put in place – sometimes at great public cost. The poor have been systematically punished (Andrews and Jacobs, 1990). Local democracy, and perhaps democratic practice as a whole, has been undermined by the introduction of state-appointed agencies such as the London Docklands Development Corporation and by the abolition of the Greater London Council and other metropolitan authorities. The land-use planning system has been partially eroded away and/or bypassed (Thornley, 1991). A systematic policy of privatisation has meant that profit-seeking organisations have moved into the provision of services that used to be the province of democratically accountable agencies. This raises a fundamental question. *Can* one move simultaneously towards the 'free market' and 'democracy'? Or, rather than being synonymous, are the two concepts divergent, even oppositional?

If there has been a move away from democratic practice – and no doubt some would dispute this – has it been in the interests of a more effective guardianship of public resources? If this was the underlying intention it does not seem to have been realised. The consistent undervaluing of the price at which major utilities have been sold has cost the public purse billions of pounds. Most of the Urban Development Corporations have sustained heavy losses on their land dealings. The policy of selling 1.5 million housing units from the public stock may have been popular electorally but, given the heavy cost of tax relief on mortgage interest, it has been extremely expensive in terms of extra cost to public funds per unit transferred. Immense capital receipts to public accounts have been forgone as a result of the heavily discounted prices at which most of the house sales occurred. The sales have also contributed to a serious decline in the amount of rented housing allocated on need-related criteria rather than by market forces. One of the main responses to this shortage has been to house families in emergency 'bed and breakfast' accommodation. The annual cost of this, if translated into capital borrowed, could have built sufficient new housing to go a long way towards solving the problem. Many of these policies have been driven by neo-liberal ideology rather than by a search for genuinely cost-effective social policy (Hamnett, 1987). They do not make a good model for the 'new Europeans' in the ex-socialist countries. It would have been better if the

Soviet Union and its dependent system of states had collapsed fifteen years earlier. They might then have had a more rational Western model to draw on.

But should the vast, complex processes that shape the built environment be answerable to democratic or user influence anyway? Does it matter? Should we worry about 'the effects of handing over the control of space, once subject to publicly accountable local government, to the balance sheet driven values of private sector retailing.' (Fisher and Worpole, 1988, 7)? It could be argued that in world terms we in Britain live in a high-quality built environment and that if the population at large were really interested in these matters they would turn out in greater numbers for local government elections. They might even make stronger representations at planning enquiries. Unfortunately such a view is based on too narrow a conception of the issues at stake. The Canary Wharf scheme, and many others that remain unlet in Docklands, is not simply a set of buildings which were built without any kind of democratic participation and which for the moment at least are surplus to market requirements. The use of finance and materials on this scale, with no commensurate return, represents a vast *financial* loss. To whom? Certainly to the developers, Olympia and York, who appear to owe something approaching £7 billion worldwide. But they can go bankrupt and to some degree protect themselves from creditors. It is the banks to whom most of the money is owed that will take the loss. They will presumably raise lending charges, make additional provisions for bad debt and to an extent restrict their future lending practices. This will make trading conditions more difficult for borrowers and may inhibit the promised business upturn – confidence in which may itself have been retarded by the property crash. In other words development episodes like Canary Wharf are not just about the view north from Greenwich, although that is not a negligible matter, but also about the destabilising consequences for financial institutions and markets. That, in turn, is about jobs, economic recovery and pensions.

The implementation of the Canary Wharf development, it should be stressed, involved little or no participation by either local people or local authority planners. It might be argued that on grounds of economic efficiency developments on this scale need to bypass the slow and costly business of citizen participation and the procedures of land-use planning and development control. Few would dispute that in the Docklands case the interests of local residents were sacrificed in a premeditated manner in order to put in place a 'state of the art' trading facility intended to keep London competitive with Paris, Frankfurt and other European cities as a financial trading centre. Given the past significance of 'invisible earnings' to the British balance of payments, and the recent shrinkage of overseas earnings by these means, one can see the point. So in today's competitive world might it be justifiable to trade off democratic procedures in the interests of economic efficiency in the renewal of the built environment? Or at least might this be justifiable for developments, such as those in London's Docklands, which

10

carry national-scale implications? This question is taken up in more detail in Part IV.

THE RELATIONSHIP OF 'MARKET', 'STATE' AND 'CITIZEN'

The Canary Wharf issue helps to highlight the three main sets of interests with a stake in the development process. These are, in simple terms, 'the market', 'the state' and 'the citizen'. These terms *are* highly simplified. 'The market' includes various different forms of capital-accumulative organisations: for example, lenders, developers, constructors, profit-driven users of buildings, and so on. Clearly their interests diverge in important ways. The producers of buildings seek high rents while commercial and industrial users seek low rents. The producers, notably the construction industry, have been unhappy for a long time with the ostensibly 'pro-market' Conservative governments of the 1980s. Gripped by a 'public spending equals bad' ideology they have failed to instigate sufficient publicly funded capital building projects to generate acceptable flows of business. 'The state' includes the central state, with its own internal conflicts between politicians and civil servants. The prime task of the nation-state is the preservation of the so-called 'national interest' – a term which can unfortunately mean anything the government of the day chooses it to mean. 'The state' also includes the local state – multi-tiered and with the same internal officer/politician tensions but with an additional sensitivity to local people and interests. 'The citizen' is the rest of us, differentially interested in, and connected to, the so-called democratic process. Typically we may be moved to some kind of organised protest action if our interests are seen to be directly threatened by a proposed motorway at the bottom of the garden. But more usually our political action is restricted to a visit to the ballot box every so often, complemented perhaps by quiescent membership of a voluntary organisation or two.

So which institutions, practices, legal frameworks, channels of communication and flows of information can best ensure that these three sets of interests, and the sub-groups within them, play an appropriate part in the process of shaping and re-shaping the urban environment? The problem is to achieve an effective balance. The former socialist countries of East and Central Europe obviously suffered from too much state and are desperately relieved to have emerged from this situation. Many in the West feel we have suffered, and continue to suffer, from too much market. Maybe the voice of the citizen can get as stifled in the latter situation as in the former. Nearly everybody, in principle, is on the side of the citizen. But competing claims are made about which particular balance of state/market activity best serves the citizen's needs. Here surely should be the focus of policy discussions. What *mix* of state and market activity is most appropriate – technically, politically,

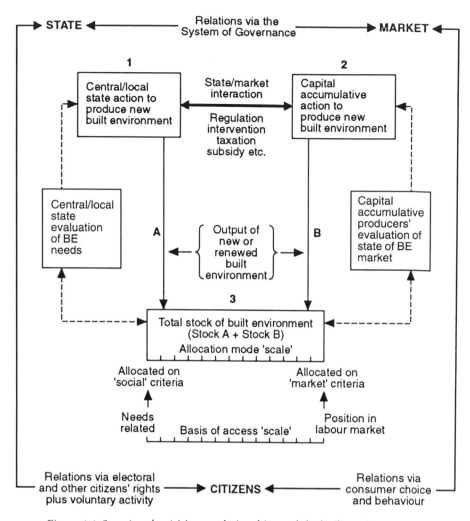

Figure 1.1 State/market/citizens relationships and the built environment.
Diagram by Peter Ambrose.

socially, and in terms of providing the necessary incentives to mobilise private capital?

Figure 1.1, which forms the basis for the more complex disaggregated model presented as Figure 3.1 in Chapter 3, may help to clarify the matter and form a framework for discussion. Consideration of the processes best begins with box 3, the total stock of buildings and other constructed environment at any given point in time. This stock is the product of hundreds of years of development activity, some of it initiated and carried out by private profit-seeking interests, some by state activity and some in other ways, for example by voluntary group or individual activity not primarily driven by the search for profit. Constant monitoring of this stock is carried out by two main sets of interests. Various central and local state institutions with statutory responsibilities – for example, housing departments, education departments, highways departments, and so on – make constant assessments of needs and initiate the construction of new buildings and facilities to help discharge these responsibilities (box 1). This produces a flow of new public construction (line A) which adds to the stock. Simultaneously profit-seeking, or capital-accumulative, organisations are making assessments of the stock based on 'market' criteria. They initiate new construction schemes (box 2) if the return on the investment to be committed looks competitive. This new private construction is then also added to the stock (line B).

The focus of this book is very much on the changing nature of the state/market interaction taking place between boxes 1 and 2. The concern is with changes in the power balance between state agencies and capital accumulative interests, because this affects the ratio of A to B in the annual output of new built environment (see the 1978–90 construction orders data presented in Table 9.1 in Chapter 9). This in turn affects the ratio of stock A to stock B. This change in the balance of the stock happens relatively slowly because the marginal addition of new built environment is only a low percentage of the existing stock in any year. But the stock ratio can also be more rapidly affected by transfers of existing property from one ownership to another, as in the case of council house sales or the sale of public utilities.

A decade or more of moves towards greater privatisation in these various relationships, and of increases in the B:A ratio of new output, has undoubtedly affected the balance of the stock very considerably. The main social implication of these changes in the ratio is that they have reduced the proportion of built environment (for example, housing) which is allocated and accessed in relation to *need* and increased the proportion that is allocated and accessed in relation to *capability to pay*. This spectrum of allocation possibilities is signified by the 'scale' at the bottom of box 3. The mass of the population too ('citizens') can be arranged on a spectrum reflecting each household's position in the labour market – or, put more simply, its income and wealth.

The key political issue is the extent to which position in the labour-market hierarchy translates directly into the capability to gain access to necessary elements in the built environment. If most of the stock of built environment were allocated at the 'social criteria' end of the scale, access would depend little on one's position in the labour-market hierarchy. But the years since the mid-1970s have arguably seen this translation become more direct. Increased production and ownership by profit-seeking interests has meant an increase in the extent to which 'market criteria' determine allocation and thus an increase too in the extent to which the user's income, rather than need, is the

13

key to access to housing, good living environment and high-standard education and healthcare. Given that these same years have seen a well-documented increase in income inequalities, it is highly likely that there has been a corresponding increase in inequality in terms of the built environment experienced. The extent of the differences in the quality of urban environments is considered with the use of census and other data in the next chapter.

The broader power relationships surrounding the production and allocation of the built environment are hinted at in the lines linking state, market and citizen in Figure 1.1. The vast majority of citizens exert influence on the activities of the state by exercising their right to vote in elections. This relationship is statutorily defined, and the electoral process closely regulated, by legislation. A much smaller proportion of people is also active in seeking to influence local and central government by means of voluntary organisations or 'pressure groups'. These vary from 'single issue' groups concerned only with a specific policy objective to those with a much broader set of concerns (see Coxall, 1990). The former are often local in scale and short-lived, and the latter national in scale and more permanent in structure.

The citizen's relationship to the array of profit-seeking producers and distributors of goods and services summed up as the market is quite different. It has some statutory content, in the form of consumer protection legislation, anti-monopolies legislation, some price control, and so on, but it is primarily a relationship based on the voluntary coming-together of buyers and sellers. Free marketeers believe that this relationship provides all necessary consumer protection, since people have choice and will buy only that which offers acceptable value for money. They argue that aggregate consumer spending behaviour will constantly exert a discipline on the market. Inefficient producers, or producers of unwanted goods, will fail. Free competition will ensure lowest possible prices. New producers will enter the market, or existing producers will diversify, to offer newly defined goods and services needs. Prices should, so far as is possible, remain 'undistorted' by subsidies. This is held by some free marketeers to be a correct strategy in moral, as well as economic, terms, since the consumer has a right to know the 'correct' price of any good or service without the distorting effect of subsidy. There are a number of problems with some of these propositions and they will be more fully explored in Chapter 8.

The third general relationship, that between state and market, has been the subject of very considerable debate since it became apparent in the early/mid-1970s that it was being redefined in certain important ways (see, for example, Cloke, 1992; Loney, et al., 1991; and Riddell, 1991). This redefinition far transcends the issue of the built environment and extends into virtually all areas of economic and social policy. The post-Keynesian orthodoxies of fiscal austerity and the perceived imperative to reduce both public-sector borrowing and the existing level of national debt have produced heavy pressures to reduce public expenditure. This in turn has placed new and far more restrictive limits on the role and range of activities of the state. In relation to the built environment, as in other areas of policy, the increasing emphasis has been on the state as the 'enabler' rather than the producer, owner and manager of facilities such as housing and public utilities (Bramley, 1993). In terms of public spending the key strategy seems to have been to reduce forward capital programmes on services such as housing while allowing revenue expenditure on, for example, Housing Benefit and emergency housing services to remain constant or even to rise. The two effects are of course logically connected. Cutting the forward building programmes that produce lower-rent housing inevitably entails a subsequent need to spend more money dealing with the resultant shortfall of low-rent accommodation. Thus the new financial orthodoxies may, over the longer term, be having effects precisely the reverse of those intended.

The state/market interface in the process of producing the urban environment consists of a mix of legislative regulation and various forms of non-legislative intervention. The legislative measures include regulations on building standards and design, fiscal measures, land-use development control, rent control and subsidy regimes. While they are 'one-way' in the sense that they are enacted and enforced by the state, they often reflect lobbying activity by private-sector organisations with particular interests, such as the 'volume' housebuilders or the 'road lobby' (see, for example, Hamer, 1987). So the route between boxes 1 and 2 is essentially a two-way one. Non-legislative intervention ranges from Planning Policy Guideline documents to less formal means of conveying ministerial intentions. It includes flows of information generated by state-funded research of various kinds. It can also take the form of obvious 'rescue' moves when the market has clearly got into acute difficulties through massive over-development. It is precisely this state/market interface that is currently at an early stage of redefinition in the case of the former socialist economies of East and Central Europe.

2

THREE URBAN
ENVIRONMENTS
COMPARED

This chapter presents a set of data, drawn largely from official sources, about three contrasted urban environments in contemporary England. They have been selected to illustrate the extreme differences that exist between the 'best' and the 'worst' local authority areas in terms of a number of quantifiable variables. They also reflect the effects of sharply varying economic processes and policy responses – or non-responses.

Wokingham in Berkshire was for most of the 1980s one of the towns that most fully typified 'Mrs Thatcher's high-tech Utopia' (see Barlow and Savage, 1991) – a town of rapidly expanding employment and population accompanied by equally rapid house-price inflation. In the 1990s it has suffered from the long-lasting recession that has hit Britain generally and the south-east in particular. The effects of this downturn in fortunes are not well reflected in the data presented, most of which are taken from the Census of April 1991. Knowsley is a metropolitan borough forming part of outer Liverpool. It includes the Kirkby residential and industrial estates. The decline in its fortunes in recent decades, and the attempts to bring about 'restructuring', have already been well discussed (Meegan, 1989). The data presented in this chapter show how limited has been the success of these efforts in the face of the massive withdrawal of private-sector investment. Tower Hamlets, a London borough, has also experienced a sharp contraction of the local economy since the early 1970s, in this case as a direct result of the dock closures that took place between 1969 and 1981. About one-third of the borough lies within the area designated as within the jurisdiction of the London Docklands Development Corporation, which was set up in 1981. The dramatic transforming effects of this element in the strategy adopted towards urban problems by the 'neo-liberal' administrations since 1979 will be discussed in Part III of the book.

First a few methodological points should be made about comparative analysis of this kind. It may be felt that little can be learned from these three areas about *regional* differences, because two of them are in the south-east region and one in the north-west. The other regions are not represented at

16

all. In fact this would be to misunderstand the nature of the problem. Differences in living standards and life chances generally occur more strikingly *within* large administrative areas than between them. When data are agglomerated and presented at the scale of the 'standard regions', the county or even the individual borough or district, much of the significance is lost and one ends up with some generalised and misleading notion of a 'deprived north' and an 'affluent south-east'. This works to conceal the difficulties of poorer or homeless people in 'rich' southern towns such as Wokingham or areas in transition such as Tower Hamlets. Many facing hardship in these areas feel doubly aggrieved by the evidence of conspicuous wealth around them. As an example, the April 1993 official unemployment rates for the standard regions of Britain varied only from 8.9 per cent in East Anglia to 14.1 per cent in Northern Ireland. But somewhat earlier data (March 1989) show that *within*, for example, the south west the rates varied by town from about 3 per cent to over 15 per cent, and in the East Midlands from 3.5 per cent to 17 per cent (Tucker, 1993).

Apart from the question of the *scale* at which data are collected and analysed, the crucial methodological question 'Disparities in what?' has to be faced. There are dozens, perhaps hundreds, of ways of measuring the quality of living conditions and life chances. To complicate matters, the data on which analysts, policy-makers and the general public seek to make judgments pass through several 'filters'. One is what might be called 'the collection filter' and it reflects the degree of commercial and political sensitivity of different kinds of data. Certain types of information about an area, for example the number of people aged 15–24 or the number of housing units built between 1919 and 1939, has little political sensitivity and is routinely collected and published. Other information, such as the level of unemployment in an area, has to be collected for a number of policy-making purposes and is also routinely published. But the level of unemployment is a very political phenomenon and the *definition* of 'unemployed' may be changed (in fact it has been – over thirty times in recent years) in order to minimise the size of the problem. 'Moving the goalposts' in this way is a politically inspired attempt to mislead. A third category of information, such as who owns the plot of land on which some new retail centre is to be developed, the balance of public to private money to be invested in the scheme, the manner in which the benefits will be shared and the annual profit or loss a certain branch plant is making, is commercially sensitive and is not routinely made available for public discussion. It comes to light only if a specially designed research project is carried out or if the issue becomes subject to a planning enquiry. Yet it is clear that this third category of information can contribute far more than the first two to an understanding of the processes that bring about and reinforce disparities between areas.

Politics, and the commercial sensitivity of investment or disinvestment decisions, are only part of the problem here. Some information is by its nature difficult to collect, collate and present in an electronically processed form. In fact in some research areas the more one relies on data that can be processed in this way the further one may be moving from the really crucial material. The essentially binary nature of electronic data processing requires a 0–1, black-and-white world. This is exactly what the world is not. The tendency to binary thinking, and the often subconscious drive to squeeze an infinitely variable reality into a finite set of machine-processable categories, often give a very limited view of urban processes and get in the way of understanding. How, for example, can the labour traditions of an area (such as London's Docklands) be quantified? Yet it is an important factor in the minds of possible investors.

QUANTITATIVE DIFFERENCES BETWEEN THE THREE AREAS

The rest of the chapter will make some comparisons between the three areas by using information collected and published by various ministries and public statistical services, including the 1991 Census and publications of the Chartered Institute of Public Finance Accountants (CIPFA). Four main categories of information will be considered. These relate to demographic structures and change, housing characteristics and costs, local occupational structures and educational provision. For most of the indicators the information is collected and made available by the relevant local authority area – the London Borough of Tower Hamlets, the Metropolitan Borough of Knowsley, and the District of Wokingham. All three are products of the reorganisation of local government which took effect in 1974. Apart from the official sources mentioned, the section will draw on the *Report on Britain's Deprived Urban Areas* published by the Policy Studies Institute (Willmott and Hutchison, 1992), which collates a number of social and economic indicators for thirty-six heavily deprived local authority areas including Tower Hamlets and Knowsley. Wokingham is not, of course, included in this study.

DEMOGRAPHIC CHARACTERISTICS

The population of the three areas for selected recent years is as follows:

Table 2.1 Population change 1961–91

	1961	1971	1981	1985	1991
				(est.)	
Tower Hamlets	205,682	165,776	142,841	147,100	158,360
Knowsley	151,389	194,095	172,957	166,300	152,091
Wokingham	62,652	99,664	115,103	135,400	138,189

Sources: 1991 Census, *County Reports*, Table B; CIPFA, *Housing Statistics*

These changes are striking. The Tower Hamlets population fell sharply between 1961 and 1981 in line with the general tendency towards the depopulation of Greater London. This tendency is now reversed, owing largely to the massive development of new housing in the southern parts of the borough since the London Docklands Development Corporation (LDDC) was set up in 1981. Unfortunately most of this housing is too expensive for local residents to buy and much of the population increase is probably due to better-off households moving in from elsewhere. The population of Knowsley borough rose during the 1960s as the industry expanded on the local industrial estates but has fallen sharply since 1971, in fact by nearly 22 per cent. Wokingham has experienced explosive growth – by over 120 per cent in the years covered in the table. As part of the buoyant Heathrow/M4/M25 economic growth zone, it is subject to continuous pressures to accept further housing development.

The 1991 age structure and ethnic characteristics of the areas (in percentages) are as follows:

Table 2.2 Age structure and ethnic characteristics

	Age structure			Ethnic characteristics	
	% 0–15	% 16–pension age	% over pension age	% white	% other
Tower Hamlets	25.6	59.2	15.2	64.4	35.6
Knowsley	24.4	60.1	15.5	99.0	1.0
Wokingham	21.3	66.7	12.0	96.7	3.3

Source: 1991 Census, *County Reports*, Tables 2 and 6

Compared to the national average, Tower Hamlets has a fairly normal age structure but a striking concentration, one of the highest in the country, of non-white people. In fact 22.9 per cent of the borough's population and over 90 per cent of some of the estates are Bangladeshi. Knowsley, by contrast, is overwhelmingly white. Older people are under-represented, perhaps because the 'overspill' movement that produced much of the growth in the 1950s and early 1960s included both new people and new jobs – many of which were for young single or recently married people who have now reached late middle age. Wokingham also has a preponderance of people of working age or younger. The reasons may be slightly different. New entrants to the area have been attracted mostly by the rapid growth of new jobs. Few of these would be for people over, say, 45, and it is certainly not an area people retire to. Nor, to judge from the ethnic data, have many Black or Asian families made their home there. Finally, data on average household size in 1991 is given below:

19

Table 2.3 Average household size

	Average no. of persons per household
Tower Hamlets	2.52
Knowsley	2.73
Wokingham	2.70

Source: 1991 Census, *County Reports*, Table 22

HOUSING CHARACTERISTICS

Information about housing tenure in percentage terms is set out below:

Table 2.4 Housing tenure

	% owned outright	*% buying*	*% private rented*	*% LA/NT*	*% Housing Assn*	*% other*
Tower Hamlets	3.8	19.4	6.9	58.3	9.3	2.3
Knowsley	13.9	39.2	3.1	39.4	2.9	1.5
Wokingham	20.9	64.3	4.5	7.0	1.0	2.3

Source: 1991 Census, *County Reports*, Table 22

Owner occupation has risen nationally (to nearly 70 per cent in the early 1990s) under the 'right to buy' policies of the 1980s – although there are now very few saleable public properties in Knowsley. The abnormal tenure pattern of Tower Hamlets has been vigorously modified by the activities of the LDDC, as we shall see in Chapter 9. Wokingham's tenure pattern is also abnormal in the 85.2 per cent proportion of owner occupancy and the extremely small proportion of rented stock. In addition, a very high proportion of these purchasers are still saddled with a mortgage and no doubt many of them have run into arrears and other difficulties as the post-1990 recession has threatened jobs and lowered real incomes. Some of these effects will be discussed in Chapter 5.

Plate 2.1 (opposite) Local authority housing in Tower Hamlets. This block forms part of the Berner Estate in an area of the borough adjoining the City. Most of the estate was built between 1949 and 1954 when the area formed part of Stepney Borough. There are over 500 housing units, nearly all flats, and the resident population is perhaps 95 per cent Bengali. The estate is heavily overcrowded, and extensions have been added to some of the blocks in recent years to help ease this problem. The block shown has not yet benefited from the Housing Authority's continuing refurbishment programme, which has been seriously affected by cuts in central government funding. By sharp contrast, several billions of pounds of public subsidy have been directed to the London Docklands Development Corporation area, whose boundary is a few hundred yards to the south. Photograph by Keith Hunt.

The three local authorities face vastly different tasks in maintaining their stock of council housing:

Table 2.5 Housing maintenance costs

	No. of LA dwellings (March 1990)	1989/90 repairs and maintenance expenditure (£)	Cost per dwelling (£)
Tower Hamlets	43,365	22,123,000	510
Knowsley	24,124	7,720,000	320
Wokingham	3,438	1,623,000	472

Source: CIPFA, *Housing Revenue Account Statistics 1989–90 Actuals*, 1991

The Wokingham housing-maintenance needs are more or less in line with national averages, but Tower Hamlets and Knowsley are facing a massive task. Many 'system built' tower blocks, a product of the government subsidy arrangements in the 1950s and the 1960s designed to encourage this form of housing (see Chapter 6), have now become extremely expensive to maintain. In recent years some of these have been pulled down and others sold. The lower-rise developments also require large-scale expenditure as a result of years of underspending on maintenance. Unfortunately comparison with earlier years shows that in both Tower Hamlets and Knowsley there has been a sharp decline in repairs and maintenance expenditure. Two years earlier these two authorities were spending £743 and £642 per dwelling respectively, on repairs and maintenance.

The incidence of both overcrowding and homelessness is very unequally spread between the three authorities. Overcrowding can be defined in a number of ways, but one good indicator is the extent to which room density exceeds one person per 'habitable room', which excludes kitchens and bathrooms. The percentage of households in this situation is as follows:

Table 2.6 Incidence of overcrowding

	% of households with more than one person per habitable room
Tower Hamlets	11.1
Knowsley	3.3
Wokingham	0.9

Source: 1991 Census, *County Reports*, Table 23

Under the homelessness legislation, families accepted as homeless by the local authority concerned must be provided with accommodation. This can be in the permanent stock of the authority or in some form of temporary accommodation. Because of the reduction in the amount of council stock available, this often means being lodged in a 'bed and breakfast' establishment – sometimes with parents and children in one room and sharing facilities with other families. This is widely acknowledged to be a most unsatisfactory arrangement which benefits no one except the proprietors of the accommodation used. Tower Hamlets, in common with many other inner London boroughs, has been driven to adopt this desperate solution on a massive scale. In fact the number of households in 'temporary accommodation' per 1000 households in the borough rose from 0.8 in 1980 to 24.6 in 1991. The corresponding 1991 figure for England as a whole was 2.5 (Willmott and Hutchison, 1992, Table 5.10).

The following table shows not only the heavy cost of these arrangements but also the sharp differences in the proportions of those accepted as homeless between the extremes of 78 per cent in Knowsley and 16 per cent in Wokingham. All housing authorities are working under the same legislation and guidelines, so the differences must reflect some combination of different degrees of housing hardship among the claimant groups and different criteria and working practices in the respective housing departments. To the extent that it is the latter, most of the explanation could well lie in the extremely small public stock available for letting in Wokingham.

Table 2.7 Homelessness and bed and breakfast costs

	Households applying as homeless	% of these accepted	No. put in bed and breakfast	1991–2 net cost of bed and breakfast
Tower Hamlets	2,212	47	1,550	£4,448,463
Knowsley	670	78	66	£18,006
Wokingham	669	16	42	£92,661
England and Wales	292,948	43	35,364	£66,430,852

Source: CIPFA, *Homelessness Statistics 1991–92 Actuals*, 1993

The net annual outlay on 'bed and breakfast' accommodation has risen nationally to over £66 million per year. As has been pointed out many times, this could be used instead to pay the annual interest charge on a borrowing of, perhaps, £750 million. This capital sum, if invested in new local authority building programmes, would produce sufficient new dwellings to house a fair proportion of those now lodged in 'bed and breakfast'. It might also serve to stimulate the hard-pressed construction industry. But such a policy would run counter to central government public-spending intentions and the drive to privatise existing housing units rather than allow additions to the public stock. It would also require a supply of suitable land. But a great deal of public-housing land has been sold off under the neo-liberal policies of the 1980s, and in the special case of Tower Hamlets considerable land in the borough has been appropriated by the LDDC to help promote speculative up-market private housing. Much of this now remains unsold, and some of the housebuilders who produced it (for example, Kentish Homes) have been bankrupted. Perhaps those who have advocated 'market-led' solutions could explain the economic and social advantages that have flowed from this combination of policies.

Local authority rents, because of the 'pooled historic cost' principle of public housing finance which still to some extent applies in public rent-setting, are much lower than private-sector rents. Moreover, they do not vary to the same extent between areas as private sector rents.

Table 2.8 Local authority rent levels 1990 and 1992

	No. of LA units (1992)	Average weekly LA rent (£)	
		April 1990	April 1992
Tower Hamlets	39,642	22.43	34.39
Knowsley	22,266	24.68	30.91
Wokingham	3,369	28.70	40.41

Source: CIPFA, *Housing Rents Statistics at April 1992*, 1992

A sizeable stock of decently maintained low-rent housing is clearly necessary in all areas where some people have low-wage jobs or are unemployed. In 1990s Britain this means in all areas. But the amount of low-rent housing, which effectively means council housing, varies enormously by area, as does the expenditure necessary to keep it in reasonable condition. 'Zone of transition' areas such as Tower Hamlets are subject to abnormal pressures. Some blocks of flats in suitable positions and condition have been sold, under centrally imposed financial regimes, to private developers for renovation and sale to City workers, or as 'second home' investments. The net result is drastically to reduce the available stock of low-rent housing. This may not matter as much in Knowsley, where the population level is declining, but it places heavy burdens on lower-income people in buoyant areas such as Wokingham or areas in rapid social transition like Tower Hamlets. Access to low-rent housing is an especial problem in Wokingham. As we have seen, the council stock is very limited, and average weekly rents in the public sector rose by 54 per cent in the four years between April 1986 and April 1990. The equivalent figure for Tower Hamlets was 28 per cent and for Knowsley 29 per cent. Between 1990 and 1992, as Table 2.8 shows, there were the further increases of 53 per cent, 25 per cent and 41 per cent in the three areas. These sharp rent rises are an inevitable outcome of the recent withdrawal of subsidies. These changes will be further discussed in Chapters 6 and 9.

The only alternative to renting is purchase. How do house prices compare in the three areas? At the time of writing (mid-1993) the housing market is still depressed, in terms both of prices and of the level of transactions, following the house-price boom that ended abruptly in 1988. The information presented is likely to be out of date by the time of publication. But the general relativities will probably remain. Further problems with comparing house-price data are that the housing stock varies by area and that there are no official statistics or indices. But free sources of data exist in estate agents' windows, local newspapers or, to be more sophisticated, from a number of building societies (notably the Nationwide Anglia). Some typical 1993 prices are listed:

Table 2.9 Typical house prices 1993

Tower Hamlets	
terraced houses	£84,000
purpose-built flats	£59,000
converted flats	£68,000
average of all properties	£69,000

Knowsley	
detached houses	£78,000
semi-detached houses	£58,000
terraced houses	£24,000
average of all properties	£62,000

(*NB* The prices of former local authority houses and flats on the Kirkby estate are considerably lower.)

Wokingham	
detached houses	£125,000
semi-detached houses	£88,000
terraced houses	£68,000
bungalows	£101,000
purpose-built flats	£60,000
converted flats	£63,000
average of all properties	£97,000

Sources: Local newspapers, brochures and Nationwide Anglia bulletins

In Tower Hamlets, the general run of housing being produced in the London Docklands Development Corporation area has little to do with local housing needs or homelessness in the borough (except that indirectly it has exacerbated the problem because some of it uses land which was formerly council-housing land). Many of the new flats and houses are not lived in or are used as second homes. This applies especially to the apartments on the river frontage, whether in converted warehouses, newly built blocks or former local authority blocks sold very cheaply to developers and subsequently 'gentrified'. Many of these have been bought and subsequently traded as speculative commodities. Others provide a 'prestige address' on London's riverside for people who might also own similar properties in Copenhagen, Amsterdam or Toronto.

In Knowsley the prices are extremely low for a combination of reasons. The population level is declining, which means that there is considerable local authority stock available, jobs are difficult to obtain, it is not an area that people migrate into, and local people usually do not have the resources to take on a mortgage. Anecdotal evidence suggests that houses can be bought for very low sums indeed on the estates of outer Liverpool.

Wokingham provides a much more typical housing scene so far as the more prosperous areas of the country are concerned. The local economy was buoyant and employment chances increasing until the advent of the recession in the late 1980s. Employees in the commercial sector can often obtain a reduced-cost loan to purchase, and many newcomers receive a 'housing relocation package' from their employers. The housebuilders have responded by concentrating on 'up-market' housing aimed at prosperous dual-career households. As the director of one major housebuilding firm remarked privately in the late 1980s, 'We can sell as many houses at a third of a million as we can build.' During the boom years the main problem for housebuilders was to obtain sufficient land with planning consent to build the quantity and type of housing being demanded by the market. Since then housebuilders have suffered from unsold stocks here as in most other parts of the country. We will come back to the issue in Chapter 5.

EMPLOYMENT STRUCTURES AND THE INCIDENCE OF UNEMPLOYMENT

The relationship of the population aged 16 and over to the labour market in each area is shown below:

Table 2.10 Employment characteristics

Area	Total population aged 16+	% full-time employed	% on govt. scheme/ unemployed	% retired
Tower Hamlets	119,921	33.8	13.5	17.7
Knowsley	115,022	30.5	14.3	17.2
Wokingham	109,548	67.2	3.3	12.0

Source: 1991 Census, *County Reports*, Table 8

The percentage of the population aged 16–24 who are engaged either on a government 'scheme' or who are unemployed varies even more dramatically by area:

26

Knowsley	30.2 per cent
Tower Hamlets	19.5 per cent
Wokingham	6.9 per cent

Source: 1991 Census, *County Reports*, Table K

It is not practicable to compare the local economies in terms of the profile of jobs available in each, since this would require a survey. For those in work the 1991 occupational structures by socio-economic group (SEG) of residents in the three areas, based on the type of employment of the main earner in each household, is shown below.

Table 2.11 Occupational characteristics

SEG	Type of job	Tower Hamlets	Knowsley	Wokingham
% 1 and 2	Employers/managers	11.4	9.4	25.0
% 3 and 4	Professional	5.4	1.8	8.9
% 5 and 6	Other non-manual	36.8	35.2	37.6
% 8, 9 and 12	Manual (higher skilled)	17.1	23.0	14.6
% 7, 10 and 11	Manual (less skilled)	27.0	28.7	11.3
% others	Others	2.3	1.7	2.6

Source: 1991 Census, *County Reports*, Table 92

Over one-third of the Wokingham main earners are in SEGs 1–4, as opposed to only one-tenth of those in Knowsley, while the proportions engaged in manual occupations vary considerably between Knowsley and Tower Hamlets on the one hand and Wokingham on the other. These patterns are explained partly by the differences of percentage of people educationally highly qualified in the three areas (10 per cent sample data):

Table 2.12 Higher-education characteristics

	Tower Hamlets	Knowsley	Wokingham
People aged 18+	10,763	10,990	10,344
% with diploma or equivalent	3.0	3.5	9.7
% with degree	7.0	1.7	12.4
% with higher degree	1.3	0.2	2.0

Source: 1991 Census, *County Reports*, Table 84

Three dates have been selected for comparison of unemployment. The first is 1983, before the deep and long-lasting recession in Western economies began, the second is 1989, and the third is 1991. The 1992 Policy Studies Institute report (Willmott and Hutchison, 1992, Tables 2.6 and 2.11) shows the following numbers of people unemployed in Tower Hamlets and Knowsley:

Table 2.13 Incidence of unemployment 1983–91

	Nos. of people unemployed		
	1983	*1989*	*1991*
Tower Hamlets	15,800	10,000	14,300
Knowsley	21,100	12,200	12,700

By October 1992 the figures were as follows:

Table 2.14 Incidence of unemployment October 1992

	No. unemployed	% rate
Tower Hamlets	16,235	11.1
Knowsley	12,375	14.9
Wokingham	3630	6.8

Source: Department of Employment, *Employment Gazette*, December 1992

Declining local economies and high rates of unemployment are likely to go hand in hand with poverty. It is not possible reliably to compare income levels by local authority area, but the proportion of the population receiving Income Support (Supplementary Benefit before 1988) can be used as a reasonably reliable surrogate. The deprived-urban-area study found the following (Willmott and Hutchison, 1992, Table 4.1):

Table 2.15 Population receiving Income Support/Supplementary Benefit

	% of estimated adult population receiving Income Support/Supplementary Benefit		
	1983–5	*1986–8*	*1989–91*
Tower Hamlets	34.7	38.3	32.6
Knowsley	32.9	34.7	33.8
England	15.9	16.5	13.9

The authors believe that the fall in percentages between 1986–8 and 1989–91 does not denote a real reversal of the trend towards increasing poverty. The first half of the latter two-year period saw relative prosperity before the unemployment figures began to rise again in mid-1990. The impact of this rise in unemployment on the Income Support figures was delayed because those losing their jobs can claim unemployment benefit for a year before becoming reliant on Income Support. In addition to this, various categories of people were excluded as claimants as a result of changes in the legislation in 1988. The real trend in worsening poverty in the two deprived areas may well, therefore, be understated. It is safe to conclude that, in general terms, one-third of the adult population in Tower Hamlets and Knowsley are living on or below the official poverty line. This is about two and a half times the national proportion. While comparable figures are not available for the incidence of Income Support in Wokingham, it can be assumed that they are better than the national average. Poverty as measured by this index therefore varies by a factor of four or more between the best-off and the worst-off areas. There are no real signs of the gap closing.

The Policy Studies Institute report sums up employment trends in Britain's deprived areas (which include Tower Hamlets and Knowsley) as follows:

> many of our selected areas of deprivation suffered even more than the rest of the country in the continuing rise in unemployment up to 1986. This was particularly true of the major conurbations.... Though all our areas shared to a considerable extent the general reduction in unemployment in the boom years 1988–90, they shared less than proportionately. Relatively speaking most of them were worse off than before.
>
> (Willmott and Hutchison, 1992, 20)

The methods of travelling to work vary considerably in the three areas:

Table 2.16 Method of travel to work

	Method of travel to work (%)			
	Bus and Underground	Car	Foot	Work at home
Tower Hamlets	44.3	23.8	18.9	2.5
Knowsley	26.0	54.8	11.7	1.5
Wokingham	10.0	72.7	6.4	4.9

Source: 1991 Census, *County Reports*, Table 82

Clearly Tower Hamlets is producing a far more environmentally friendly pattern of commuting than is Wokingham. In view of the considerable debate about the growth of 'home working', which might be expected to be growing fast in a 'high tech.' area like the M4 corridor, it is worth noting that the percentage of Wokingham people working at home is still only 4.9 (just 0.9 per cent higher than in 1981).

Striking differences in the rates of car availability help to explain the commuting patterns:

Table 2.17 Rates of car ownership

	No. of households	% with no car	% with 1 car	% with 2+ cars	Cars per household
Tower Hamlets	62,844	61.2	31.9	6.9	0.46
Knowsley	55,750	52.1	36.7	11.6	0.61
Wokingham	50,904	11.4	40.1	48.5	1.46

Source: 1991 Census, *County Reports*, Table 21

The very high proportion of two-car households in Wokingham probably owes a great deal to the 'company car'. Since 1991, it is possible that car-ownership rates have risen still further in areas like Wokingham (to judge from the growing congestion on the M25/M4).

EDUCATIONAL PROVISION

It is very difficult to make meaningful comparisons between the standard of educational provision in different areas. Simply to compare average standards of achievement in school-leaving examinations is to miss a lot of the point. Pupils in different areas have differing extra-school conditions to contend with. In areas such as Wokingham there are likely to be far more parental assumptions about educational achievement than is the case in areas like Tower Hamlets or Kirkby. The extent to which conditions at home are conducive to study varies enormously, as we have seen, as does the quality of recreation available and the sense of security as young people grow up. These are all important factors bearing on the motivation and capability to study effectively.

None of these contextual factors appeared to carry any weight when the government decided in 1992 to publish the notorious 'league tables' of levels of attainment for each school in the country in terms of success in the public examinations for the GCSE and A/AS levels. These appeared as a set of 108 booklets, one for each local education authority. The results for the three areas under review were as follows:

Table 2.18 Secondary educational attainment

	% with 5+ GCSE at grades A–C	% with 1+ GCSE at grades A–C	% taking an A/AS level	Average A/AS score
Tower Hamlets	16.0	53.8	56.9	5.6
Knowsley	16.3	44.8	62.7	7.9
Wokingham*	47.0	74.0	95.5	15.5
England	38.1	65.5	86.1	15.0

*Unweighted average for the four Wokingham schools (abstracted from the Berkshire data).
Source: Department for Education, *School Performance Tables*, 1992

These differences are very striking, especially in terms of the low percentages taking an A level in the two poorer areas and the very low A-level scores achieved compared to those of Wokingham. Clearly the chances of going on to further and higher education are very unequally spread. It is easy to understand why people in Tower Hamlets and Knowsley may come to believe that this opportunity is not for them, especially if language barriers and culturally based expectations about gender roles add further complications. To reduce the inhibiting effects of such attitudes positive discrimination in terms of educational spending is required. The following figures show that, to some extent, this is happening – but clearly not to the extent necessary to overcome the home, environmental, cultural and perhaps language difficulties that have produced the unequal outcomes seen above.

Table 2.19 Per capita educational expenditure

	1990/1 net institutional expenditure per pupil (£)	
	Nursery and primary	Secondary
Tower Hamlets	1939	2491
Knowsley	1338	2209
Berkshire (for Wokingham)	1267	1876
England and Wales	1300	1964

Source: CIPFA, *Education Statistics 1990–91 Actuals*, 1992

Using the same source, we can see that this pattern of spending produces considerable differences in pupil:teacher ratios:

Table 2.20 Pupil:teacher ratios

	Pupil:teacher ratios		
	Nursery and primary	*Secondary*	*Total*
Tower Hamlets	17.4	14.1	16.0
Knowsley	22.0	15.0	17.4
Berkshire (for Wokingham)	23.3	15.6	19.2
England and Wales	22.2	15.5	18.8

Source: as previous table

The two poorer areas, and especially Tower Hamlets, are apparently managing to allocate above-average amounts of funding to produce favourable pupil:teacher ratios. But the data given earlier on attainment levels indicate that much more compensatory investment is required in the more deprived areas in order to make the best use of the abilities of those living there.

Finally, some indicative data on tertiary education expenditure in 1989/90 are shown below:

Table 2.21 Mandatory higher-education grants

	Mandatory awards expenditure (£)	*June 1990 population (estimated)*	*Expenditure per capita (£)*
Tower Hamlets	1,214,000	166,900	7.27
Knowsley	3,414,000	157,400	21.69
Berkshire	23,163,000	755,500	30.66

Source: as previous table

This completes the story evident from the 'league tables' data. Young people living in Berkshire are roughly four times more likely than those in Tower Hamlets to reach the attainment level that entitles them to a mandatory grant to develop their abilities in post-secondary education.

QUESTIONS RAISED BY THE DATA

This chapter has presented a range of evidence to describe disparities in living standards between two of the poorest and one of the richest local authority areas in Britain. The task of collecting and presenting comparative data in this form is difficult and time-consuming. Even so the analysis is necessarily superficial because certain factors bearing on the ease and comfort of life are difficult to evaluate. There is no obvious way of showing how supportive it might be, in an aggressively alien land, for ethnic minority groups in Tower Hamlets to live on estates peopled largely by those of similar language, culture, traditions and religion – even if the physical conditions are bad. Nor can we assess how stressful it might be to live in Wokingham for the poorly paid or unemployed when public-sector rents are rising rapidly and house-price levels preclude purchase. In Knowsley there is no obvious way of measuring the way people feel about the long-term economic decline of a once proud and buoyant area. For following up such issues one would need area-based statistics which are not readily available – for example, comparative data on tranquilliser prescriptions per person per year.

The comparative material does, however, prompt a number of questions which provide part of the agenda for the rest of the book. These questions fall under three headings: analytical, ethical, and political. The analyst pure and simple (if such a person exists) seeks to understand how the phenomena dealt with have come about. How do patterns of investment and disinvestment occur? By what processes do they turn into measurable differences in the factors which condition life-styles and life-chances? Are these differences decreasing or increasing? Do changes in these factors, and the effects they have, seem to lead to more or less social tension and economic efficiency?

The related ethical questions cannot easily be ignored. *How much* disparity in living conditions is too much? Are widely disparate, and often diverging, rates of unemployment, homelessness, hypothermia or some other adverse condition acceptable or not? Is it morally justifiable to reduce direct taxation if the consequence is reduced spending on welfare and on educational and housing provision for those largely dependent on state-provided services? Is the close interpenetration of great wealth and great poverty in an area like Tower Hamlets tolerable in the late twentieth century in one of the richest and most technologically advanced countries of the world? Is any of this a problem?

Finally, whatever one's ethical judgments, there are the political and administrative questions. What social and economic policies seem to produce what distributional outcomes? Are the outcomes those intended by the policy-makers? Who *are* the policy-makers, and whose interests are they serving? How do we exert influence on whoever they are? Is public policy-making synonymous with wielding power to produce change, or are key decisions being made outside the democratic sphere? What changes are occurring in the power relations of central and local government, and between the deprived and the affluent? Will market forces rule, and the state have withdrawn completely from the urban arena, by the year 2000? And to what kinds of politics might increasing inequality give rise? As this book goes to press the candidate of the explicitly racist British National Party was elected to Tower Hamlets council from Millwall Ward in the Isle of Dogs – the first electoral win by this party. Many people feel that extreme politics of this kind can flourish where social injustice is manifest; and perhaps nowhere is this more evident than in this area of east London. Maybe we ignore these trends at our peril.

Consideration of the data leads to the tentative conclusion that further withdrawal of state financial support and regulation could well have the effect of increasing levels of inequality. It is striking that on certain indicators which we could call Set A – such as the level of council rents and pupil:teacher ratios – the areas do not differ that significantly from each other or the national average. The same is true of standardised mortality ratios (roughly, the rate of deaths taking population structure into account). In 1990 these were 103 for Tower Hamlets and 108 for Knowsley (England and Wales = 100) and both areas had moved towards the national average since 1986 (Willmott and Hutchison, 1992, Table 4.7). But on other indicators – for example, unemployment rates, house prices and degrees of dependence on Income Support: call them Set B – the areas varied dramatically by factors of up to three or four. In the case of the Set A indicators it is largely the state, in the form of long-established and publicly funded housing, education and health-provision systems, that is at work. But variations in the Set B indicators, the supply and quality of jobs, income levels (which together condition dependence on Income Support) and house prices, are primarily produced by the interplay of market forces. It seems reasonable to conclude that the further incursion of market-driven organisations into the provision of employment and the delivery of housing, education and healthcare may lead to future differentials between localities that are more characteristic of the Set B than the Set A indicators.

These questions concerning the evident decline in democratically based intervention in the processes that condition life chances and experiences, in other words the state:market ratio of power and activity, are important. We shall return to them in the next chapter by attempting to assess how the power balance is changing between democratically accountable and non-democratically accountable agencies in the fashioning of the built environment, and later on in Chapter 9 we will consider some of the apparent results of these changes.

Part II

PROCESS

How the urban environment is fashioned

This part of the book seeks to explain the processes that generate and re-generate the urban environment. The intention is to clarify the sequence of stages in the production of the most common elements in the built landscape – homes, offices, shops, warehouses, factories, leisure centres, and so on. Since the buildings that currently make up the urban environment were mostly built sometime during this century, the four chapters each have a historical dimension to show how the particular production process has changed over that period.

Chapter 3 presents a general model including the stages of promotion, financing, construction, allocation and subsequent management. The various agencies and institutions influential at each of these stages are identified and divided into those that are formally democratically accountable and those that are not. The aims and strategies of the actors in each of these two categories are considered.

Chapter 4 deals with private development undertaken as investment where the aim is capital accumulation via a future flow of rentals. Normally in present-day Britain these processes produce 'commercial' development – offices, shops and industrial premises.

Chapter 5 deals with another category of development undertaken for capital accumulative motives – property built for sale. In Britain this almost always takes the form of housing built speculatively for owner occupation.

Chapter 6 discusses a type of development undertaken by public agencies – housing built and managed by elected local authorities. It considers the political situation that led to this public intervention in the previously 'free-market' processes that produced new built environment and discusses the ways in which this intervention is changing.

Chapter 7 considers development undertaken by what is known as the 'voluntary sector'. This is predominantly housing produced and managed by a wide variety of organisations that share a number of characteristics –

'non-profit' intentions, 'voluntary' management and partial independence from local political control. This form of development is, generally speaking, more common in other European countries but currently enjoys growing political and financial support here.

3

THE SYSTEM GENERATING NEW BUILT ENVIRONMENT

The processes that produce additions to and renewals of the built environment are initiated and carried out by three main sets of interests – profit-seeking private investors, legally defined public authorities, and 'voluntary' organisations, groups or individuals. Nearly all initiators, or promoters, of built forms are readily classifiable into one of these groups, although definitional problems can arise. The aim of this chapter is to present a model which disaggregates the production process, identifies the various types of organisations and institutions engaged in it and differentiates those that are democratically accountable from those that are not. A later chapter (Chapter 9) assesses the negative change that occurred over the period 1978–90 in the ratio of democratic to non-democratic input into the process. That chapter goes on to analyse changes in the 'mix' of output of new built environment produced over this period of de-democratisation – roughly the period during which 'neo-liberal' Conservative administrations held power.

MOTIVATIONS

It is as well to consider first the range of reasons for the initiation of new development, because there will be little chance of understanding the processes and outcomes unless motivations are understood. Attention to motivations will also reinforce the point made in Chapter 1 that it is not some disembodied 'society' or 'natural forces' that are at work here, but real interests and people with very clear intentions and strategies.

The profit-seeking promoters have, by definition, the motive of capital accumulation. They are bound to be guided in this way because commercial practice and company law alike require that their main concern must be the interests of their shareholders and creditors. Profit-seeking promoters undertake additions or renewals to the built environment simply as a special case of the general accumulation process. This requires that an initial capital sum (M) is committed to the purchase of a number of production commodities (C). These include building materials, labour in various forms and a site. These inputs are then combined in a production process (P) in order to produce a finished commodity, in this case a building or other built form (C '). The value of this commodity is then realised on the market as M' (see Ambrose, 1986, 1–8 for a fuller discussion). The quantity M' – M, if positive, represents the gross profit which is then subject to taxation. The result is the net profit. If M' – M is negative, a loss has occurred on this particular scheme. Since rates of profit are time-related, the crucial issue for an investor is the anticipated value of net M' – M per time period (T), which is normally the accounting period of one year.

Land development is different from the capital accumulation process for the general run of commodities, such as saucepans or word-processors, in at least three main ways. First, a piece of land is necessary, and this site forms part of the value of the finished product, mostly because it has a future redevelopment value. Second, the finished product cannot be moved to the market but remains where it was built. Finally it has a 'semi-permanent' life – perhaps on average 80–150 years in the case of most buildings. All these special characteristics carry implications. The value of the underlying site outlives the building and forms an increasingly important element in the financial valuation as the possibility of redevelopment draws closer. The immobility means that the market value of the building will be related to local events which stand outside the production process itself, for example trends in employment or changes in communications facilities in the area where it was built. The semi-permanence means that the promoter has a choice of two main ways in which the value M' can be realised. She/He can retain the ownership and benefit from the future stream of rents (see Chapter 4) or sell the product on completion and realise M' as a capital sum (see Chapter 5). In practice numerous combinations of these two realisation forms are possible.

Public authorities and organisations have for most of their history operated on quite different motivations from profit-seeking promoters. As the complex structure of British local government built up from its modern origins in the mid-nineteenth century it became the main 'deliverer' of social policy (see, for example, Bruce, 1968; Jones, 1991; and Dearlove and Saunders, 1991, chapter 10). Local authorities, elected by universal franchise, were given the tasks of providing and managing a wide range of activities, from social services and firefighting to libraries and refuse collection. Many of these services, for example the provision of housing and roads and the exercise of land-use control, have played a very significant part in the evolution of the built environment (see Chapter 6). These services have been financed partly by funds from central government and partly by taxes levied on local property or, during the period of the ill-fated Poll Tax, on every local elector. Investment and management decisions relating to each activity of a local council are taken by a committee made up of elected councillors. These committees and the council are advised by local government officers with professional qualifications who are appointed by the council on a permanent basis.

The motivations of the councillors and appointed officers engaged in activities that shape and reshape the built environment, for example by carrying out housing or redevelopment schemes, are quite different from those of managers of the same processes in the profit-seeking sector. While not invariably as pure as the driven snow, motivations are normally financially disinterested and based upon some notion of 'public service'. The councillors have, or should have, no financial incentive in the activity since they earn their livelihood elsewhere while the officers' incomes do not directly depend on the profitability of the decisions they are making. In fact 'profitability' in strict accountancy terms is not a very useful concept in many areas of public-service provision. The value of publicly promoted developments such as housing, community centres or libraries can be judged only in terms of some qualitative assessment of the social utility or public benefit produced. It cannot be measured in terms of return on investment committed, since benefit of this nature is not adequately reflected in monetary units. This does not, however, remove the constraint that all publicly managed activities must make the most cost-effective use possible of public resources. Overall, therefore, motivations in the public sector are different from, and less clearcut than, those in the accumulative sector.

The motivations in the voluntary sector, whose development activities are dealt with in Chapter 7, are more difficult to summarise. Typically a voluntary sector agency such as a housing association comes into being because a self-selected set of people feel that there is an identifiable group whose housing needs are not being met. The voluntary agency is able to promote new construction, or convert some existing building, because appropriate advisory programmes and funding sources have been made available as part of central state policy and with various forms of help and support from the local authority. Thus the state has legislated an enabling framework in which local individual and group effort can produce environment-modifying activities. Sometimes such organisations have grown to enormous size and appear not dissimilar to commercial organisations in their behaviour. Even these, however, operate nominally on a 'non-profit' basis. Thus their motives are not primarily accumulative and they seek to provide housing at the lowest possible rents consistent with acceptable quality. In these respects they resemble local authorities. But they are different from local authorities in that their existence, and the range of functions they carry out, is not statutorily required or defined. They are also different in that their directors and managers are not elected and so are not, in this sense, democratically accountable.

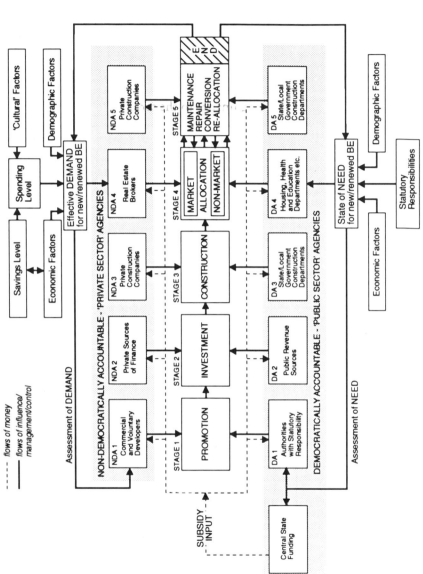

Figure 3.1 The system producing and maintaining the built environment. Diagram by Peter Ambrose.

HOW IS THE BUILT ENVIRONMENT MODIFIED?

The production and subsequent life of any element in the built environment can be divided into the five discrete and consecutive stages shown as 1–5 in Figure 3.1. The stages (with stages of the accumulation cycle in brackets) are:

1 PROMOTION – making the decision to produce the building or facility in the chosen location and taking the first steps to implement the decision

2 INVESTMENT – committing finance to the scheme (M in the accumulation cycle) in order to assemble the inputs (C)

3 CONSTRUCTION – combining the inputs, including land, materials and labour (P), in order to construct the building or facility (C')

4 ALLOCATION – making the building or facility available to the first user on one of a number of possible bases (producing either M' or social utility in some form)

5 SUBSEQUENT MANAGEMENT – maintaining, repairing, perhaps reallocating and/or converting the building or facility until it reaches the end of its usable life (all of which can produce further M' or social utility).

The upper and lower sets of boxes set out typical agencies that may be operating at each stage but there is no necessary one-to-one relationship between agency and stage. The entire process may be sequentially integrated so that a single organisation carries out or manages all five stages. This appears to be very rare, since even in previous socialist systems such as the German Democratic Republic a local municipality would depend on central state funding for finance and contract the local building *Kombinat* for the construction. Perhaps the purest case of sequential integration would be the self-builder of a house who initiates the scheme, uses her/his own savings, does the construction work her/himself and subsequently lives in and maintains the property. A British private speculative builder normally organises stages 1–4 but often depends upon inputs of loan finance at stage 2 and perhaps an estate agent at stage 4. Since the product is disposed of at stage 4 the builder has little interest in stage 5 (unless seriously bound by a warranty). An investment company, by contrast, may enter the chain at stage 4 by acquiring a property on the market and then managing it for part or all of stage 5. In this case M is the purchase cost translated into the annual cost of servicing the loan taken out for the purchase and M' the annual stream of net rents. The arrows showing reciprocal movements between stages 4 and 5 indicate that the property may change ownership by reallocation and/or may be converted to a different use during its lifetime.

The 'feedback' arrows from 'End' in stage 5 back to stage 1 show that new promotion decisions depend partly on the rate at which existing buildings or facilities are judged to have reached the end of their usable life. 'Usable' may be defined in terms either of commercial criteria or of social utility. An older house that is nearing the limits of habitability may still provide useful shelter for those in dire need, maybe for squatters, or it may be that further investment would give it some additional life as a permanent home. If so its continued existence would provide further social utility. Or an old, technically obsolete, office building may be providing a flow of rents that is only a fraction of those that would result from a new office block on the site. If older houses stand in the way of some potential redevelopment for a use that will provide higher rates of return, as frequently happens in 'inner urban' areas, commercial criteria may well dictate that they should be demolished even though, as physical structures, they may have years of life in them. Similarly London office blocks built in the 1960s are even now being demolished to make way for blocks with a higher technical specification. Whether or not 'End' has been reached therefore often hinges on financial and/or political, rather than technical, considerations. In more theoretical terms, the issue may well turn on whether the interests controlling the process are working in terms of 'use values' or 'exchange values' (see, for example, Harvey, 1989, 100–5).

A central aim of the analysis in this chapter, and of the book, is to explore the relationship between formal democratic structures or other forms of citizen empowerment and the evolution of the built environment. To further this exploration the organisations and institutions that produce and manage the events at each stage have been grouped into two categories which are shown in the two horizontal sets of boxes above (NDA 1–5) and below (DA 1–5) the central set. From the brief description in each box it should be clear what, in reality, these organisations and institutions are. Conventionally the agencies operating to produce or renew the built environment are divided into 'private' and 'public' agencies – that is to say, 'market' and 'state' agencies. For certain purposes this is a useful duality. But when analysing environment-providing processes it is too crude.

For example, can the London Docklands Development Corporation (LDDC), a housing association, a self-builder and The Housing Finance Corporation be categorised as either 'state' or 'market'? All are hybrid forms, just as the provision process itself is almost invariably hybrid. For example British council housing is, or increasingly was, promoted by a local authority, funded largely by 'private-sector' loans, built mostly by profit-seeking contractors, allocated by the local authority and subsequently managed by that authority unless sold or conveyed to a voluntary organisation such as a housing association. Such a form of development is therefore dependent on both sectors at various stages. It could not, in total, accurately be classified as either 'public-sector' or 'private-sector'. There is a further problem with the simple binary division. Many institutions such as British Rail, the Port of London Authority and so on are still nominally, or in terms of ownership, 'public'. But such organisations have for several decades now had the duty laid upon them to achieve specified rates of return on capital employed or have in some other way been obliged to act as if their main objective were the accumulation of capital rather than the provision of a 'public service'. These two objectives may well be regarded as synonymous by neo-liberals (see Chapter 8), who tend to regard service provision and profit-making as invariably coincident activities. Those who do not see things this way are left with the problem of sensibly categorising institutions such as those mentioned.

The present analysis, therefore, steps around the distinctions 'private and public', 'market and state' and categorises agencies active in the provision of built environment in terms of *the ways in which their existence and their power to act are legitimated*. The test is the simple one of whether or not those with power in each agency were put in place by electoral means for a specified time period and thus whether they can be removed by these means when that time period is up. In other words the distinction is whether the electorate, as a whole or locally, had rights in the process conferring power. This test removes most ambiguities from the categorisation process. The agencies in the shaded area above the central boxes are 'non-democratically accountable' (NDA), and those in the lower shaded area are 'democratically accountable' (DA). It should be stressed at this point that no value judgments have been made about the degree to which DA agencies may, or may not, be more sensitive to user needs or convey greater or lesser degrees of empowerment to people than do NDA agencies. One of the aims of the book is an attempt to clarify judgment on this issue – in other words to encourage consideration of the question 'Is the ballot box a sufficient, or even the best, way to ensure genuine power-sharing?'

The question of motivation and accountability has already been touched on but it requires more detailed discussion. NDA agencies can be subject to different forms of accountability and thus are driven by a range of motivations. Private developers, financial institutions, construction companies, estate agents and property professionals such as architects and engineers are normally accountable to a set of shareholders, depositors or partners. Any member of the public can become one of these providing she/he has the skills to do so or the money to deposit savings or acquire shares. Thus the capability to affect the behaviour of the production system in these cases is accessed by skill or money. Other NDA agencies, for example the LDDC, are run by small groups of people appointed directly by a minister of state or a local government department to which they are accountable. People affected by the actions of such agencies have no direct power to affect their behaviour unless they themselves become members of the appointed group. A third set of NDA agencies are 'voluntary' – they come into being simply because people want to form them. These include housing associations, co-operatives and people carrying out self-build or self-promotion activities either individually or in groups. Most of these may, in fact normally do, take steps to enlist user participation in the environment-generating activities they carry out, but they are not democratic under the strict definition adopted.

DA agencies, by contrast, normally have a statutorily defined status together with a set of powers and duties prescribed by law. They include the central and local government authorities responsible for housing, transport facilities and the provision of services such as healthcare and education. All these responsibilities require them to act at stage 1, and perhaps at later stages, in the development of new built environment. It is a characteristic of DA agencies that they respond to the feedback information link labelled 'Assessment of need' in Figure 3.1. They have little or no interest in acting as land developers for reasons other than to produce buildings and facilities that help them to discharge their responsibilities. By contrast, most NDA agencies, except perhaps the 'voluntary' ones, act more in accordance with an 'Assessment of Demand' feedback link. In recent decades this 'market-led' behaviour has also become increasingly characteristic of organisations such as British Rail, the Port of London Authority and even the Church Commissioners, which, under less strict criteria, would be regarded as 'public'.

Stages 1 and 4 have especial significance to the shaping of the built environment and the ways in which people achieve access to the buildings and facilities they need. The level of promotional activity at stage 1 largely conditions the rate at which the environment is renewed. Not only that, but the interests involved at this stage normally maintain some control over later stages 'downstream'. Construction schemes are promoted by a wide range of interests. Whether NDA or DA, they have to act within the democratically managed framework of the land-use planning system. In Britain this system still has certain powers, despite the long retreat from the more interventionary system initiated after the Second World War (see Ambrose, 1986; Brindley, Rydin and Stoker, 1989; Thornley, 1991; and Rydin, 1993). But the powers have always been largely negative in the sense that the planning authority's main role is simply to confer or deny permission to initiate a construction scheme in the context provided by a previously agreed land-use development plan for the area. The local public authority, whether county or district, therefore has *reactive* power over proposals for development which takes place characteristically on privately owned land. This form of intervention is much weaker than the *proactive* power to shape development patterns that would derive from the financial capacity to initiate development on a major scale. This in turn is weaker than the power that is derived in other systems from the local authority's role as the main supplier of land for the development process. In these latter systems the local municipality has significant negotiating power with the private sector constructor in the process of shaping the pattern of new development (see Chapter 10 for an account of the Swedish land-development arrangements).

Stage 4 is of crucial social significance since it conditions the mode of access to the new or renewed elements in the built environment. From the point of view of sensitivity to the general pattern of needs it might not matter if stages 1–3 and 5 were dominated by NDA organisations so long as a reasonable proportion of the environment produced, in the forms of housing, shops, open spaces and so on, were allocated in stage 4 by DA agencies working on social criteria rather than solely on capacity to pay (a distinction set out also in Figure 1.1). If market criteria become the sole or even the primary basis of access, then, as argued in Chapter 1, the income and wealth inequalities already existing in society will to that extent be reproduced in the capability to gain access to the built environment. This, in turn, will work over generational timescales to reinforce and amplify societal inequalities. Conversely, a reduction of these inequalities can be, and clearly has been, achieved by allocating certain elements added to the built environment, for example council housing and public open space, on 'non-market' criteria at stage 4 – in other words by making allocations on the basis of assessed need rather than capability to pay.

In practice, especially in relation to housing, a large element of non-market allocation in stage 4 is very likely to mean that profit-seeking organisations 'upstream' from this stage see insufficient financial incentive to initiate new promotion. This will be so unless sufficient public subsidy is provided at stages 2 and/or 3 to render the M' – M increment less dependent on the allocation stage, or unless public money is put into users' pockets so that they constitute a more promising market at stage 4. The source of subsidy support is the box marked 'Central State Funding'. In Britain this means the Treasury, although in practice the funds are channelled via the major spending departments of state, and some of the money is ultimately spent via local authorities as they discharge their statutory or other responsibilities. The various ways in which public support can be applied to the land-development process are shown in Figure 3.1 by the dotted lines leading to all the ten boxes representing the agencies and organisations active in the process. This issue of public support is so crucial that the matter deserves to be put more succinctly. There's no profit in carrying out development for the poor unless the process is subsidised. So two policy questions arise: how much public support is justifiable and how should it be applied?

The first of these two questions depends on the state of the economy, the spending priorities of the government of the day, the pressures applied by international funding agencies, and about a thousand and one other variables. It can fairly be described as an imponderable. But the second question of policy is well worth consideration, in the interests of efficiency in the use of public money, no matter what the overall level of support. The first issue is the balance of support between that applied to the 'private sector' (the lines to NDA 1–5) and that applied to the 'public sector' (the lines to DA 1–5).

We shall see later on (Chapter 8) that some neo-liberals assert that public funds channelled to the DA sector are consumed in some magical way and 'produce nothing'. This can safely be disregarded as ideologically inspired nonsense. But the issue of the *balance* between public funds spent on, for example, offering incentives in the form of cheap loans to private construction companies and those spent subsidising a local authority is a genuine one in any mixed economy.

A separate but related policy issue is the stage at which the support can most cost-effectively be applied. There are sound arguments for providing state support for 'production' (most commonly via stages 2 and 3) rather than for 'consumption' (stages 4 and 5). In the case of housing, consumption support into users' pockets at stage 4, for example Mortgage Interest Tax Relief (MITR), is likely to feed through into higher prices and rents in a manner that benefits various categories of producer. In particular (see Chapter 5) it is likely to inflate land prices and thus benefit land vendors. This means that it is largely self-defeating in terms of assistance to housing users. This is almost inevitably bound to be so unless effective controls are simultaneously imposed on the two main forms of housing-access payment – prices and rents. In British housing history rent control has had a mixed and highly contested effect since its introduction in 1915 and price controls have never been attempted. It is also likely that state support in the shape of production-stage incentives will have a bigger effect in stimulating rates of output. This will in turn, given constant demand, both reduce prices and increase construction companies' turnover. But the strategy of applying support at the first three stages of the housing-provision process is at odds with current neo-liberal orthodoxies. These hold that most support should be available at stage 4 to enable everyone to compete 'fairly' in the 'free market'. This is in sharp contrast to the strategy for motorway and other road construction, where support is directed to the producers and the user is not directly charged at the point of 'consumption' of the road space.

Stage 4 is important for another reason. Promoting agencies, whether led by a 'market' or a 'non-market' logic, are guided in their actions by the two feedback loops already identified. These do not lead directly from the rate at which the built environment is reaching 'End' but pass through the 'filters' of 'Effective demand' and 'State of need' respectively. To take the first of these, effective demand is conditioned by demographic factors such as population level, age structure and household size – although to some extent the last of these is itself dependent on the number of housing units accessible. The state of demand at any given time is also conditioned by economic factors such as income levels and past savings levels. The rate of interest on borrowed money, the capital sum available as a loan for any given income level and the extent of the state's fiscal concession to purchasers or renters of property are also, as has been seen in recent years, very significant in setting demand levels. Finally cultural factors such as societal norms about the proportion of disposable income one might be expected to spend on accommodation or transport, as distinct from other forms of consumption, will affect the changing assessments of demand. The assessments of need made by public authority promoters take into account a more limited range of factors, for example savings levels are less significant, but they too depend on a range of demographic factors and on the range of statutory responsibilities to be discharged by the public authority.

The model of the built-environment provision system presented in this chapter therefore not only disaggregates the various discrete stages of the process and identifies key 'players' at each stage, but it also highlights a number of the central policy issues to be considered when analysing the evolution of the environment. For example does the social significance of each stage vary? If so, one implication is that it matters little if the promotional effort, the financing and the organisation of construction are undertaken by organisations largely beyond electoral accountability, so long as the product of the process, new housing or other necessary facilities, is allocated in some democratically sensitive fashion. But how effectively, anyway, is 'user sensitivity' achieved by means of electorally based democracy? Does casting a vote periodically in a local election mean that citizens are fully empowered? Or, in relation especially to the shaping of the built environment, are better methods of empowerment available? In a broader sense, is it justifiable that we should all get the standard of built environment we can afford to pay for, as would be the case if stage 4 were managed by NDA agencies on 'market' criteria? Or are larger issues at stake? Subsequent chapters will take up some of these questions.

4

PROFIT-SEEKING
DEVELOPMENT
– AS INVESTMENT

The key point about the mode of development to be dealt with in this chapter is that the promoter and the provider of finance (stages 1 and 2 in Figure 3.1) are initiating a scheme to produce something over which one or other or both will retain some longer-term interest. The intention is to benefit from a future stream of rentals. In more precise terms, they are seeking the maximum possible net annual rate of return on the effort and funds invested – 'net' because the profit (the $M' - M$ increment) is the total annual income from the scheme less the total annual outgoings. They are also seeking to ensure that the 'yield' from the scheme – the net income expressed as a percentage of the current capital value – is fully competitive when compared with other investment possibilities. It is after all the capital value that, theoretically at least, can be turned into a different form of revenue-earning asset such as government stock, shares or money in a bank deposit. In the past decade there have been a number of commercial schemes built for immediate sale – a reversal of previous trends. These are not dealt with separately in this chapter because the change of ownership soon after completion does not change their essential nature as investment developments or modify significantly the process by which they are produced.

Development for investment at the present time is primarily directed towards producing buildings that have a 'commercial' purpose – as offices, shops, industrial premises, warehouses, and so on. This is easy to understand because the driving force behind this form of development is the search for rents. By contrast to housing rents, which are limited by household incomes plus any targeted housing-allowance support, commercial rents have no inherent built-in ceiling. They form a variable part of the cost structure of activities carried out by entrepreneurs who may well cope with high rent increases by cutting costs elsewhere in their operations. Some commercial activities can by their nature tolerate much higher rents than others. For example a diamond dealer can turn over millions of pounds of business in a very small office which can be on any floor of a building. She/He can afford to pay extremely high rents per square foot of space used. Other activities, for example selling furniture, demand more space per business transaction and are more conveniently carried out at ground-floor level. The affordable rent will be lower. These differences help to explain how different retail uses 'sort themselves out' in urban areas. Commercial rents also tend to have a closer relationship with general business cycles than do housing rents. In fact, as later data in this chapter will show, the amplitude of their fluctuations is often greater than that of the general business cycle. This is partly because they have rarely been subject to any statutory limitation, whereas residential rents in Britain have been under the 'dampening' effect of some form of control almost continuously since 1915. So while before that date the development of housing to rent had been commonplace it has now become only a small part of the investment development scene.

In any investment choice there is always an opportunity cost. Effort and funds could well be invested in alternative ways such as in stocks and shares, foreign currencies, government securities, commodities, Rembrandts, or even farmland with development potential. Investment decisions always involve trading off risk against reward. A risk is not precisely predictable or it would not be a risk. Reward is sometimes predictable (for example in National Savings issues) but perhaps more often is not. Generally speaking the higher the risk taken by the investor, for example a bet that Gillingham will win the FA Cup, the higher the reward expected – maybe a return of 5000 to 1 if it comes off. At the other extreme one might help to fund the government's spending programmes by investing in 'gilts' or Treasury stock. One might pay £1000 for a piece of paper issued by the government in 1994 that promises to pay back £1000 in 2007 and 8.5 per cent of this sum per year in the meantime. In this case virtually all the risk and reward variables are known in advance – except that the rate of interest promised might be above or below rates actually achievable over the period. This latter variable will affect the value of the piece of paper, which can, of course, be traded in the period before the redemption date. There is a vast range of investment possibilities between these two extremes both in terms of risk:reward ratios and, very often, of the timespan over which the investment can be expected to pay off.

THE NATURE OF PROPERTY
AS AN INVESTMENT

Property investment has its own particular characteristics. In this mode of development the promoter and the financier, and the agents involved in lettings and management, have a long-term interest in the building or complex. The constructor or building contractor usually has, by contrast, a shorter-term interest. Once the construction contract has been completed, and the payment made, the builder's interest in the scheme is confined to whatever maintenance work needs to be done under the terms of the agreement with the developer.

By nature, a completed development has certain inherent disadvantages compared to other forms of investment, and to summarise them is to summarise the risks (for a detailed discussion of risk assessment in the land-development process, see Byrne and Cadman, 1984). The scheme has been designed for one purpose, perhaps as an out-of-town shopping centre, and often cannot easily be adapted for another. It has been built in one location and cannot be moved to another. It has been built in one technological era and may find its marketability destroyed in another – for example canals are not currently profitable. It may depend on patterns of behaviour, such as playing squash or going to the cinema, which change through time. It needs constant maintenance and management. It is vulnerable to the undermining effects of subsequent, therefore more up-to-date, developments for the same purpose which may be built nearby. Its accessibility may be reduced or enhanced by transport developments in the area. Even if all other factors remain constant, government fiscal or general economic policy may undermine the net returns from the scheme. Finally, it is a highly illiquid form of capital asset and may be difficult to dispose of quickly if it becomes unprofitable. Thus the rewards from the scheme depend upon a flow of rentals which in turn depend on demand for this type of space, in this location served by this transport network, in the anticipated states of technology, user choice and behaviour, and governmental regulatory regime. This is a formidable set of risks to assess for the scheme's lifespan, which may be fifty years or more.

To set against the risks there are a number of advantages in favour of property as a form of investment. Although rents for facilities such as shops and offices may fluctuate considerably in the short term, the underlying long-term tendency is for commercial rents to rise in current money terms as the economy grows and as inflation occurs. The capital costs of putting the scheme in place, by contrast, are fixed at one point in time – when the loan was taken out to cover the construction and associated costs. The subsequent loan-servicing costs tend to decline in real terms as a proportion of the income from the scheme. Thus the overall effect is often to provide a good protection against inflation. Furthermore at a certain point the loan taken out to cover construction costs may be paid off entirely although the scheme may have many years of 'earning' life remaining. As the earning capacity rises in terms of current money, so the 'book' value of the scheme, some multiple of the net annual revenue, will rise also. This will enable the owner of the building to borrow more funds on the security of the rising value and thus carry out more development. Apart from these anticipated $M' - M$ gains in terms of net rentals, the scheme is an asset which may be sold. This produces a final $M' - M$ increment as a capital sum and a termination of the promoter's interest. So, at least in most 'normal' states of the property market the

50

promoters and financiers of investment schemes have options – they can remain the owners, collect rents and borrow on the collateral of the property to start more developments, or sell out and use the money on other investment activities, whether property or otherwise, elsewhere.

DEVELOPMENT FOR INVESTMENT
– SOME HISTORY

Property schemes as a form of investment go back a long way. In fact until well into the present century this form of development probably predominated over all others – which is why this chapter precedes the next three. Imperial Rome had its private rented-housing sector and no doubt the promoters/investors/managers of the properties tried then as now to pack in as many rent-payers as the density regulations allowed and to skimp on repairs expenditure. Medieval merchants invested some of their trading profits in property schemes. Probably many of these were for ordinary people to rent, but some money must have gone to fund prestige projects such as the development of formal town squares or the building of a cathedral. The latter gave little hope of a realistic rental return, but the funders probably looked to dividends on a higher plane.

More recently investment in rental-housing development was prevalent in Europe as the Industrial Revolution began to induce mass urbanisation. Evidence comes from diverse sources. As a successful pianist and composer in Vienna from the early 1790s to his death in 1827, Beethoven lived at over forty addresses in the private rented sector in and around the city (see Forbes, 1973) – which might be regarded as a reasonable sample of the accommodation the sector had to offer. Typically he occupied a flat in a four- or five-storey block – often on the upper floors, which was not good news for piano shifters. He was a difficult tenant, increasingly noisy on the piano as his deafness got worse and with undesirable habits such as pouring water over his head (and the floor) and doing calculations on the walls. All his friends and cronies, as distinct from his titled patrons, seem to have been tenants too. Clearly only the very rich could buy property.

The owners of most of these rented properties appear to have been members either of the small group of Austro-Hungarian aristocrats, for example the Baron Pasqualati or the Esterhazys, or of the Vienna business community. As examples of the latter, one of Beethoven's landlords was a master baker, one a bookbinder and another a lawyer. At other times he lodged in a house owned by the Augustinian order. So clearly the ownership of the sector was to a degree fragmented. But there were also some very large mixed-use developments. The Trattnerhof on the Graben, a main road leading down to St Stephen's Cathedral, was developed by Johann Trattner, who had made a fortune in publishing. It accommodated about 600 people on the upper floors and had a number of shops at street level (Braunbehrens,

Plate 4.1 Terraced housing built primarily for renting. This particular development in Nunhead, south London, was built in the late 1870s and is more fully discussed in the text. The rapid urban expansion on this part of London's urban fringe was no doubt related to the easy access to central London given by three nearby railway lines built in the previous decade. Little has changed externally in a century or so of use, and various attractive decorative features survive, although the iron railings and gate were probably lost in the Second World War. The reliance on open-fire heating is shown by the number of chimneys shared with the house next door. Photograph by Keith Hunt.

Plate 4.2 The Marine Gate flats on the Brighton seafront. This block of 132 luxury flats was completed in 1939. At that time rents for unfurnished flats ranged from £140 per year (one bedroom) to £450 (four bedrooms) – a sum for which a three-bedroom semi could be purchased. There were garages, enclosed gardens, a restaurant and bar, and room-service meals. Flats were available on three-, five- or seven-year leases, but only half were rented when the war started, the remainder being used to accommodate naval officers. In 1971 the block was offered for auction sale as an investment producing gross about £75,000 per year – an average rent of about £575 per unit per year. Most of the residents, working through a company, purchased the headlease in 1975 and finally in 1986 the freehold. The block is now effectively self-managed by the residents. Photograph by Peter Ambrose.

1991). The development was reputed to bring in more revenue annually than a small principality – in fact these large developments were known as 'stone earldoms'. There was also, it seems, a steep rental gradient sloping 'down' from the centre. Opera-lovers may care to know that Mozart wrote *The Marriage of Figaro* in an expensive city-centre flat and *Cosi fan tutte* in a cheaper flat in the suburbs – not that one can tell from the music.

During the nineteenth century London, and other British cities, expanded rapidly with housing developed largely for investment purposes. The process has been expertly analysed in a study of Camberwell, a suburb in south London (Dyos, 1961). Development was carried out largely on building leases granted by the freeholders of the land. Typical landowners were the landed gentry, such as the de Crespigny family, and bodies such as the Haberdashers Company and the Corporation of London. In this particular case the railway companies had opened up new areas for development in the 1860s by improving the access to central London (the house illustrated as Plate 4.1 is within ten minutes' walk of three railway stations). Builders, the vast majority of them building only five to ten houses per year, bid for the leases on the land. These were granted typically for some period between sixty and ninety-nine years and often with covenants which might, for example, specify the quality or type of housing to be built and the provision of facilities such as pubs and churches. The ground rents agreed might be geared so as to 'deal in' the freeholder to the future value to be created on the land or, in the case of less well-advised freeholders, might undervalue the future rental income the lessee was to obtain.

Builders obtained capital from building societies which worked at a very localised scale (for example the Lambeth Building Society). Typically the loan might be for fourteen years at an interest rate of 5 per cent or so. Other sources of finance included lawyers or other well-off individuals. The rate of construction was very rapid at various times in the century, especially in the 'boom' years of the later 1870s. In fact in the absence of any overall assessment of demand over-building sometimes occurred, and there are reports of 40 per cent of the new houses standing empty in one local survey in the early 1880s (Dyos, 1961, 82). This is an early example of the characteristic supply-side 'lumpiness' to be discussed more fully later in the chapter. The demand was very much for single-family dwellings rather than flats. Builders used a number of designs from standard source books (for 'mansions', 'villas', 'lodges', etc.), on which they often superimposed their own embellishments. Thus houses in the expanding suburbs often had bay windows, perhaps on two floors, decorative plaster work around the bays and porches, carefully tiled paths and iron railings to reinforce the 'Englishman's home is his castle' aspect. The builder, normally the head lessee, sometimes sought to sell the houses leasehold either to the occupier or to another investor. In this way he could pay off loans, minimise interest charges and fund himself for further developments. But the vast majority of

Plate 4.3 Part of the 'shopping parade' development built to serve the retailing needs of the speculative 1930s Patcham owner-occupancy estate (some of the 1300 houses of which are illustrated in Chapter 5). In all there are eighteen shop units, all with flats above. This parade of eight, not all visible, reflects very well the changing pattern of retailing needs of the residential population, many of whom are elderly. There is a ladies' dress shop/haberdashery, a video lending library, a dispensing chemists, an off-licence, a vacant unit which was previously an estate agent, a butcher's, a bookmaker's and a bank. Two of these eight units are on the market. Photograph by Peter Ambrose.

the new houses were retained for the rentals so that the flow of income covered, hopefully by a good margin, the ground-rent and financing costs.

Coming closer to present-day Britain, changing social, economic and technological conditions have produced changes in demand for different types of investment property schemes and a general retreat from developing housing to rent (Kemp, 1993). In the 1930s, shopping parades to service the rapidly expanding suburban estates were an attractive proposition. This was especially true if the shops were adjacent to a railway or underground station, which would itself act as a stimulus to residential development in the area. Many of the new commuters would need to do some shopping on their way home in the evening (see Jackson, 1973; Rose, 1985). In the inter-war years there was also a brief wave of building purpose-built flats to rent, partly to meet the arguments of conservationists concerned at the rate at which farmland was being built over on the urban fringes. These developments used the new concrete-frame techniques, were often 'modernist' in style and were designed to give as much natural light to the interior as possible by the use of, for example, central wells or cruciform shapes. But rents could not compete with mortgage repayments. One authority (Burnett, 1991, 273)

Plate 4.4 Part of Hollingbury industrial estate on the northern fringe of Brighton. Started in the late 1940s, this estate was intended to improve the light manufacturing base of a town which has always had a heavy service-sector emphasis. Some of the buildings were classics of their time and have even been characterised as 'poor man's Bauhaus'. Activity grew during the prosperous 1960s, and by 1971 there were 5000 manufacturing jobs on the estate. Now many businesses have closed down. The buildings are mostly unsuitable for today's capital-intensive manufacturing and there is inadequate parking space. In 1993 there were only 800 industrial jobs surviving, and much of the space had been occupied by retailers such as Asda, MFI and Carpetright, and American Express has converted an old factory for office use. Photograph by Peter Ambrose.

quotes the 1930s rents of Pullman Court in Streatham as between £68 and £130 per year and of Embassy Court in Brighton as from £120 to £500. Such blocks, with their garages, restaurants and sporting amenities, were clearly aimed at 'professional and business' households, preferably childless. Meanwhile Wates, for example, were building houses for sale to people *earning* only £150–200 per year and paying only 7s 6d per week (or about £20 per year) on mortgage repayments (Ambrose, 1986, 17).

Since the Second World War investment property schemes have been developed to carry out three main types of use – office activities, retailing activities and industrial activities (see Marriott, 1967, for a racy and fur-lined account of the development scene in the early post-war years). All three user markets have changed considerably in terms of the total annual demand for new built space, the quality, design and location of the buildings required, and the rents which, characteristically, the users can afford to pay for the space. These changes reflect developments in the form and nature of the activity in question – for example new business technologies call for high-specification buildings, and many 'high-tech' industrial activities require buildings quite different from traditional factories. In addition new categories of development, for example retail warehouses and office parks, have come into being to serve new patterns of trading and business.

THE VOLATILITY OF THE INVESTMENT DEVELOPMENT MARKET

The aggregate demand, the supply and the resultant rent levels for the three main types of investment development have fluctuated quite considerably over the past three decades. These fluctuations, which form one of the inherent risks for investment-property entrepreneurs, flow from a number of characteristics of this particular accumulative process. The most striking of these is that the rate of new promotion tends to be 'lumpy'. This supply-side variability has been extensively discussed at a theoretical level (see, for example, Harvey, 1982, chapter 8, and 1985) as a characteristic outcome of unregulated private investment behaviour. At a more empirical level it seems to reflect a pervasive optimism, held widely but not collaboratively, about the future state of the economy and the future demand for particular kinds of space. It has sometimes also reflected changes in overall economic policy such as periods of credit liberalisation. Unfortunately the 'gestation period' for a major development, which may be five years or more, is quite long enough for both market conditions and government policy to have moved against the scheme between promotion and completion.

To take the office market in Britain as an example, if the annual average level of aggregate demand for space for the years between 1962 and 1987 is given the index value 100, the demand level in specific years rose in a fairly regular manner from about 53 in 1962 to nearly 200 in 1987 (Key *et al.*, 1990). But the supply characteristics were far more volatile, owing partly to an ill-advised attempt by government in the mid-1960s to limit the amount of office development taking place in London and elsewhere in the south-east (the 'Brown Ban'). Thus the supply index fell from around 100 in 1964 to under 50 in the late 1960s, rose sharply to around 150 at the height of the 1972–3 property boom, fell again to late 1960s levels in 1975, and then rose sharply but irregularly to about 230 by 1987.

Inevitably, in the absence of any regulation, this interaction of a relatively steady rise in the demand curve with the highly fluctuating supply character-istic led to considerable instability in rents. These showed an abnormally high level in the early 1970s then a long spell below the period average level during the 1980s. The steady rent rises in the period 1968–71, followed by a sharp rise in 1972, triggered a massive supply-side increase from 1970 to 1973 as finance institutions and developers rushed into new schemes to cash in. This had the effect of 'glutting' the market, and rents fell sharply, in index terms, from 1973 to 1977. But because of their long development period many of the office schemes which were started in the period of rapidly rising rents became available for letting during this time, thus depressing rents still further. It should be stressed that these index figures are for the country as a whole. Particular local centres such as the City of London or a provincial centre such as Leicester could well show these effects to a more marked degree. Similarly a new and highly speculative office zone with previously unknown rent characteristics, such as London Docklands, might well have been expected to show extreme fluctuations in both supply and rent levels – which is exactly what it has done (see Jones Lang Wootton, *50 Centres*).

The market for retail developments has shown a similar volatility. The demand for shopping space has also been on a steadily rising trend, but the supply has varied much more sharply around the mean. The result has been that rents have varied from about 60 per cent of the period average to about 145 per cent. In this category too there was a sharp increase in supply in 1971–3, 'tracking' a rental increase, followed by a drastic fall in both rent levels and supply which lasted to 1977. The period 1983–7 saw an unprecedented increase in the supply of retail premises. Much of this new retail space has failed to find tenants at the expected rent levels, or even any tenants at all, as the business recession of the early 1990s has affected retail turnover and thus capacity to take on new premises and pay rising rents. This effect has naturally proved a disincentive to the promoters of new retail schemes and this will have the effect of decreasing supply. But this adjustment will come too late to save many investors suffering from drastically falling rents.

The market in premises for industrial use has shown quite different characteristics over the 1962–87 period. This has been a period of dramatic change in a number of key aspects, such as Britain's share of world manufacturing output and trade, and the amount and type of labour required to achieve a given volume of output. Industrial processes too have changed dramatically, with the end of the 'smoky chimneys' era and the growth of cleaner 'high-tech' industrial processes. These developments produce changes in the design and specification requirements of the buildings in which industrial activities are carried out (see Morley *et al.*, 1989; Fothergill, Monk and Perry, 1987). Changes in distributional techniques have also generated a steady demand for warehouses, an increasingly important sub-category of industrial premises (Debenham Tewson Research, 1993a). In the industrial-premises market it is again the demand index that has remained the steadiest over the period, ranging from about 80 per cent of the period average to about 112 per cent. Supply has been on a generally falling curve, with spectacular short-term variations, while rental growth over the period has been negligible, with a long-term fall since the high point of the early 1970s. These aggregate figures, again, mask very significant local variations and especially the generally reduced attractiveness of certain traditional manufacturing areas in the inner cities and the growth of cleaner industrial

Plate 4.5 A typical retailing/office investment development from the 1960s property boom –
the Whitgift Centre in Croydon. This town saw a massive expansion of office space as a
business 'overspill' area for central London in the early 1960s. The story is well told in Marriott
(1967). New design features of the shopping centre include complete pedestrianisation, two
levels of small retail units, and some degree of weather protection for window shoppers. The
shops and services on offer include a Woolworth's, a giftshop, a travel agent, secretarial services,
men's clothes shops and a shop called 'Ooh La La' which could be selling anything.
Photograph: Topham Picture Source.

activities on the urban fringes where they benefit from the greater accessibility conferred by the growth of orbital motorways.

Apart from these three main categories of investment development there are a number of others, most of them with a shorter history (the definitions given are those adopted by Hillier Parker, a firm specialising in property analysis). In recent years there has been a rapid growth of 'office parks' (see also Debenham Tewson Research, 1990a). These are complexes situated outside towns and often in landscaped or park-like surroundings. They have at least 10,000 square feet of developed space in buildings of two or more storeys. They are typically situated adjacent to motorways or other main roads and normally have one parking space per employee. 'Retail warehouses' have mushroomed in the 1980s, often in suburban or urban fringe locations, where several of them may be found on adjacent sites. They are typically built using simple modern technologies and have a one-storey floor area of at least 30,000 square feet. They retail a range of commodities which require extensive display space such as carpets, furniture and 'do-it-yourself' goods. Several of these buildings often occur together to form a 'retail park'. Since these too depend on car-borne shoppers they have considerable adjacent parking space. Another recent form of development has been the out-of-town regional shopping centre. These are large complexes, with 500,000 square feet or more of space. By the end of the 1980s such centres included Brent Cross in north London, Merry Hill near Dudley in the West Midlands, and the MetroCentre outside Gateshead. At least twenty others have been proposed or are under construction but given the general economic climate of the early 1990s it is likely that many of these will be re-scheduled for later completion.

Since these new forms of commercial development have a shorter history it is not possible to chart their long-term fortunes. Data from Hillier Parker show that the rental index for office parks had moved down to 92 by the end of 1992 (May 1989 = 100) although there was considerable regional variation. For example the index was 158 for the East Midlands but only 78 for East Anglia. The rent index for retail warehouses, by contrast, has increased steadily in all regions since May 1986 (100) to, for example, 175 in the East Midlands and 164 in East Anglia (Hillier Parker, *Specialised Property Summaries*). Since this has been a period of rapid growth of supply of both these forms of development it is clear that demand, at least for retail warehouse space, has been very buoyant in recent years. The sharp business recession of 1990 onwards is, however, likely to have serious effects on demand for developments of both kinds, especially in the south-east.

FACTORS INFLUENCING DEMAND AND SUPPLY

The factors underlying these trends in the demand, supply and rental characteristics in the various markets for investment property reflect the deep-seated long-term changes occurring in the economic and social organisation of the society as a whole. Most non-residential buildings are built to accommodate people at their work. The following table illustrates some of the changes that have occurred in the pattern of employment in recent decades.

Table 4.1 UK Employment changes 1971–90 (000s employees)

	1971	1990
All employment	22,139	22,855
(male)	13,726	12,050
(female)	8,413	10,806
Manufacturing	6,783	4,422
Distribution, catering and repairs	3,686	4,824
Banking, finance, insurance, etc.	1,336	2,734
Other services	5,049	6,936

Source: HMSO, *Social Trends 1992*, 1992, Table 4.10

Employment in manufacturing has fallen to only 65 per cent of the 1971 figure. Fortunately the changes in the nature of much British manufacturing, from heavy to 'high-tech' light industry, has served to maintain some level of demand for new industrial premises. By contrast the number of workers in the three 'services' groups identified has each risen very sharply – in the case of financial services the number has more than doubled. All these activities generate a demand for very specific forms of property – warehouses, modern retailing developments, offices of various types and very often mixed developments incorporating several of these.

Data on trends in household expenditure also help to explain the long-term changes in the demand for retail space of various kinds.

Plate 4.6 (opposite) A 1962 artist's impression of the proposed 35-storey BP House to be built on a site immediately north of London Wall, the boundary of the 'square mile' of the City of London. This was to provide accommodation for 2000 office workers and to be built with reinforced concrete walls clad in stainless steel with curtain walling units between. The view, looking westward along London Wall, shows the striking contrast between this 'state of the art' development, and similar blocks in the mid-distance, and the four- and five-storey late nineteenth-century office buildings on the left. The 1960s blocks are themselves now technically obsolete and were ripe for redevelopment in the 1980s office-building boom. How long might their successors last?
Photograph: Topham Picture Source.

Table 4.2 UK household-expenditure trends 1976–90

1985 = 100		
	1976	*1990*
Food	96	104
Clothing and footwear	68	115
Household durables	83	120
Catering (meals etc.)	90	150

Source: HMSO, *Social Trends 1992*, 1992, Table 6.2

Food expenditure has not risen substantially in real terms although rising rates of car and home-freezer ownership have affected the demand for retail facilities in terms both of location and of design. The suburban, out-of-town or regional-scale shopping centres now attract an increased proportion of total trade compared to the corner shop or conventional shopping street (Debenham Tewson Research, 1990b). These changes have generated a demand for such developments. But the future trajectory of that demand depends upon a wide variety of factors, including car-ownership rates, fuel costs, 'green belt' policies and planning and conservation restrictions in inner-city areas. The increases in expenditure on goods other than food, for example clothing, footwear, furniture and other goods for the home, have been very considerable and these too have contributed to the demand for new retail space appropriate for trading in these items.

As we have seen, these changes in demand for different types of built space are relatively orderly and predictable compared to the extreme fluctuations of the supply response. This supply variability has had an inevitably de-stabilising effect on the trends in rentals. This creates a problem in that medium-term unpredictability in rent levels achievable complicates the promoter's financial calculations and increases the risk taken by lenders. It may be as well to try to characterise the various 'eras' in the investment-property history of Britain over the past thirty years and to seek tentatively to identify the causes of the supply-side variability. Some of this reflects the collective behaviour of the main investing institutions. The 1960s was a period of steadily rising rents, in real terms, in the three main user sectors. Property development therefore appeared as a relatively reliable and stable avenue of investment, and a considerable flow of money was attracted, especially from the financial institutions with a long-term investment horizon – the pension funds and life-assurance houses. The general stability was enhanced by the relatively firm supply control exerted by the planning system, although the attempt to exercise stringent control over the rate of office development in London in the mid-1960s may well have led to a rush to develop when the controls were removed, leading in turn to over-supply problems.

The early 1970s saw a new liberalisation of financial policy under the

Plate 4.7 Two large commercial developments in the West India Docks area of London's Docklands. That on the left is South Quay Three. This is a fifteen-floor development with a total of 207,000 square feet of high quality office space to let. The publicity material points out that space is available at £10 per square foot (mid-1993) compared to £35 in the West End of London, that rates are only 35 per cent of those elsewhere and there are 210 'secure and clamp-free' parking spaces. Total occupancy costs per square foot are therefore £17.42 compared to £58.36 in the West End. The open-plan interior can be flexibly partitioned, energy costs are minimised by a Building Management System and all space is carpeted and air-conditioned. Finally the Jubilee Line extension, when completed, will mean a ten-minute journey to the City. Despite all this it is not yet fully let.
Photograph by Keith Hunt.

'Competition and Credit Control' measures introduced by Anthony Barber as Chancellor. The aim was to increase the amount of money available from banks and other institutions so as to increase investment in Britain's increasingly outdated manufacturing sector. The money supply increased by about 25 per cent from mid-1971 to mid-1972. But much of this new lending was directed towards the apparently secure haven of property development or towards the purchase of imports. Institutional investment in property schemes more than doubled between 1972 and 1973, and it continued to rise in real terms through the mid-1970s (Ambrose and Colenutt, 1975; Rose, 1985, 163). The result was a huge surge in the supply of property. The damaging effects of this instability in the property market were sufficient for the government to appoint an advisory group to report on the matter (Advisory Group on Commercial Property Development, 1975). Arguably the rush to develop was driven as much by investment funds seeking a hedge against inflation as by any rational assessment of the demand trends. These were more or less flat for retail and industrial uses and only modestly upward for office space. The lack of buoyancy in demand was hardly surprising. The quadrupling of oil prices following the 1973 Middle East war helped to

63

Plate 4.8 The Little London office/retail scheme in Mill Street. A complex of former warehouses on the south bank of the Thames just downstream from Tower Bridge has been converted into office space, shop units and a restaurant, totalling over 75,000 square feet. The scheme includes an enclosed courtyard, underground parking, 24-hour security and underfloor cabling facilities. The office accommodation is available (as at 1993) on five-year leases, with the rent rising from £5 per square foot in year 1 to £13 in year 5. Previously this area, adjacent to St Saviour's Dock, contained a mixture of nineteenth-century warehouses, granaries, and grain and seed mills. Photograph by Keith Hunt.

provoke a widespread recession in Western economies in 1974. World trade contracted, inflation multiplied, real returns on property investment fell or even became negative, and the property slump of the early 1970s was well under way.

The election of the first Thatcher government in 1979 saw a marked swing towards deregulation in financial markets, as in nearly all others. The early 1980s saw a sharp increase in the availability of credit of all kinds, and the economy expanded rapidly in the mid-1980s. At this period it seemed as if the expansion in activities such as financial services and currency dealing would continue indefinitely. It seemed that the main precondition necessary to ensure London's continuing primacy as a financial centre was the right kind of 'state of the art' buildings to provide the physical accommodation for these activities. These buildings needed, among other requirements, much greater inter-floor space than hitherto to allow for the cabling and ducting necessary to service the array of modern business machinery. Office blocks dating from only twenty years previously were now seen as obsolete in the face of the new specifications.

In the case of the City of London there was a special factor compounding the supply-side volatility. The government's heavy financial and other support for the redevelopment of the Docklands area, channelled through the LDDC, posed a severe threat to the more established business areas of London. Clearly there were sites in the Isle of Dogs for literally millions of square feet of new office construction – in fact there was space within the LDDC designated area as a whole to build as much commercial space as currently existed in the City. This new space could be produced to the latest specifications. Yet given the amount of subsidy the government seemed willing to direct to the area, for ideological as much as for more rational reasons many would argue, this space was likely to come on to the market at far lower rents than could be offered in the City or Mayfair.

The Corporation of London's response, led by the chair of a key committee and the chief planning officer, was to shift to a much more liberal planning stance and actively to encourage the promotion of a number of very large schemes in the City. The Corporation sold 'air rights' over roads and railway lines to facilitate larger developments, gave very rapid responses to 'prestige' applications and appeared to view proposed developments in largely architectural terms and without reference to any master plan that might have sought to equate projections of demand with the scale of the supply. The result was a dramatic transformation of the City skyline with schemes such as One America Square, Alban Gate, Minster Court and Broadgate – which alone includes about 4 million square feet of office space. It seems very unlikely that, collectively, this set of newly developed space is achieving a volume of rentals that would justify its production in financial terms. The situation has arisen because one burst of deregulated supply-side activity, in Docklands, triggered another, in the City. It would be interesting to see exactly how the losses are being spread.

In various other ways the pattern of events in commercial property investment in the mid- to late 1980s was quite different from that of the previous boom in the early 1970s. Then the financial institutions had provided most funding for development so that by 1980 land and property represented 26 per cent of their total assets (Central Statistical Office, *Financial Statistics, 1965–1990*). In the 1980s expansion of activity it was the banks who provided most of the finance. In fact at the height of the boom the banks had a total of £40 billion lent out to property companies. The institutions, perhaps influenced by the low-yield performance of the property sector in the early 1980s, tended to 'sit out' the boom. By 1989 property represented only 13.6 per cent of their total assets. At the time of writing this seems to have been a very sensible decision.

Another difference lay in the internationalisation of financial interest. Direct property investment from overseas rose in current prices from £80 million in 1981 to £2 billion in 1988 and over £3 billion in both 1989 and 1990, before falling to £1.2 billion in 1992 (Debenham Tewson Research, 1993b). There have been rapid changes in the source of this influx of funding. Swedish and Japanese funds were active early on, in the latter case using UK balances earned by the surplus in trade. These were followed by Dutch and French investors. By 1992 over 57 per cent of the overseas money came from German institutions, with the Middle East as the next most important source. The Japanese and Swedish inflow had shrunk to almost nothing. These rapid changes in the sources of funding illustrate how easily international capital can switch between world markets and how inherently volatile property finance has become. Taking the period as a whole, about 75 per cent of the inward investment was concentrated on London, and about the same proportion was attracted by office schemes. It has been observed by several experienced practitioners that some of these overseas investors knew perhaps less than they should have done about the volatility of the British, and especially the London, property market.

The attraction of London, at least for the time being, lies in the high rents still achievable even after two years of rapid decline since the late 1980s:

Table 4.3 Prime office rentals in various European cities

	ECUs per square metre per year (1992)	% change 1990–1	% change 1991–2
Brussels	210	+6.3	+3.9
Copenhagen	121	0.0	−6.6
Paris	609	−8.3	−6.0
Berlin	495	+33.3	+6.7
Frankfurt	530	+5.9	0.0
Amsterdam	218	+22.2	−11.9
Madrid	480	−8.3	−11.9
Stockholm	324	−13.8	−5.3
Geneva	491	0.0	−13.1
London	560	−26.9	−27.1

Source: International Commercial Property Associates, *Europe in Perspective, Rents and Yields*, September 1992

It is, incidentally, interesting to note the rapid rise of rents in Berlin following German reunification. This is also evident from an overview of European commercial property markets as at 1992 produced by Debenham Tewson and their European partners (Debenham Jean Thouard Zadelhoff, 1992).

A third difference from the early 1970s boom in London was that many of the developments of the late 1980s were built for disposal once tenanted, or once the first rent review had been reached, rather than retained for the long-term interest in the stream of rents. Many were purchased by overseas investors seeking to benefit from London's continuing significance (with New York and Tokyo) as one of the world's three great financial centres (see Budd and Whimster, 1992). In part this new strategy probably reflects the nature of the predominant funding source. As a general rule banks seek to make shorter-term loans, and the value of the developments they fund needs to be realised by a sale so that the loan can be repaid. Pension funds and insurance companies characteristically attract longer-term deposits in the form of pension contributions and life-policy premiums. This means they are in a better position to 'lend long', thus enabling the developer to realise the value M' over a longer period.

THE ELEMENTS IN A PROPERTY SCHEME

The implementation of an investment property scheme requires the combination of a number of elements. These are a site, appropriate finance, a workable design, planning consent, a building contractor and, on completion but preferably before, successful marketing of the space so as to produce the stream of rental income. It is the developer's task to find and combine these elements in a way that maximises the rate of return on the capital invested.

The developer

The old-style buccaneering property developer so graphically depicted in earlier literature (for example Marriott, 1967) may have dominated the market up to the early 1960s but he is no longer a significant player in the market. The 'seat-of-the-pants' methods of site evaluation and development appraisal would not succeed in today's more sophisticated scene. Research and the analysis of time-series data sets play an increasingly important role in property-investment decisions and risk evaluation (see, for example, Byrne and Cadman, 1984; Healey and Nabarro, 1990; Property Market Analysis, 1992). The long boom that began in the mid-1960s saw property development emerge as one of the leading activities in the whole economy and it proved to be the birth period of the modern property industry. In the period up to the beginning of the slump in late 1973 the process was largely investment-led and the major investors were the institutions. But the banking system, in the guise of the 'secondary' banks, also became heavily involved as lenders. In the 1973–4 crash, when rents and capital values fell sharply, many developers were unable to keep up capital and interest payments. This led creditor institutions to take over the equity and control of some development companies and to play a more active role in development themselves. It also led to the failure of several secondary banks (see Rose, 1985).

The early 1980s was a period of relative standstill in the property world and by the mid-1980s there was a shortage of commercial space to service the economic boom which was occurring and which looked set to continue into the 1990s. The later 1980s saw, therefore, another property boom – but with some different players. The older and more established developers such as Land Securities and Hammersons tended to continue their pattern of building to retain the ownership of the future rental stream – in other words to continue as investment developers. They also, by virtue of their longer period of operation, often owned both completed developments and landbanks. They were thus able to benefit from redevelopment possibilities on sites in their ownership.

But much of the new development was carried out by a small number of 'trader developers', including companies such as Stanhope and Speyhawk. Their main strategy in the boom was to carry out an active development policy, using the readily available bank finance, and then to dispose of the tenanted schemes by freehold sale. They also traded actively in land. Much of their interest was in big prestige schemes such as Broadgate in London which were designed with close attention to the occupier's needs. These needs too were becoming more sophisticated as the technologies of business dealing and financial trading became ever more complex and more demanding of 'intelligent buildings'. It was often these developments that attracted the interest of foreign investors in British property. Much of this activity by the trader developers was 'highly geared' – that is, it was based heavily on loans rather than new capital raised by selling shares in the company. It was therefore extremely sensitive to the increases in the rate of interest on borrowings which occurred in the late 1980s. This degree of exposure to both repayment schedules and interest charges has caused major difficulties to several of the trader developers, especially since rents have fallen sharply (see Table 4.3). One major company in this category, Rosehaugh, is now in liquidation.

The site

Development of any kind requires a site. Generally speaking, investment-property schemes occur in one of three or four location types. The most prestigious schemes are office or retail/office schemes occurring on 'prime' sites in or near the centres of major cities – in the case of London, in the City or the West End. A second possibility lies in the redevelopment of older retail centres either in suburbs long established as shopping centres or in medium-sized towns in the provinces – often termed 'secondary sites'. Often in these cases there is the chance to redevelop a covered shopping mall with offices on upper floors in the 'backland' area between two central roads or to pedestrianise one or more streets. A third possibility is a retail complex or business park on or near a peripheral bypass or motorway, especially where it intersects with a major radial route. In this case calculations may have been made about the number of potential users or employees who live within, say, a thirty-minute car journey. This number may run into millions in densely populated parts of the country and this should ensure long-term growth in retail turnover and other forms of business. A fourth characteristic location is in 'windfall' areas, where, for example, the 'air space' in a major railway terminus such as Victoria can be filled with a retail/office development which will draw business from the many thousands using the station daily. Or former dock, industrial or airport land can be used for any of the three major investment-development purposes. Examples of 'windfall' land from former airport uses would be Croydon in London, and Tempelhof in Berlin.

Finance

A very large proportion of development for investment purposes is carried out on borrowed money, which is why the promotion stage is differentiated from the financing stage in Figure 3.1. Thus activities are heavily influenced by conditions in the capital market. The post-war period has seen a continual growth in activity by institutional investors so that they now dominate nearly all investment markets. The annual amount of savings handled by these institutions grew from £7 billion in the mid-1950s to £115 billion in 1981 and over £300 billion by the late 1980s (Healey and Nabarro, 1990).

The modern era of investment-property development dates from the early 1960s, when profitability conditions began to appear highly promising. For example in the London office market rents had inflated at a yearly compound rate of 8 per cent from 1954 to 1964. For the ensuing six years this annual growth rose to 15 per cent (Rose, 1985, 160) – owing in no small part to the supply constraints resulting from the restrictions on development imposed by government. These conditions interested especially the insurance companies in the late 1960s and early 1970s. Property investment by these companies and the pension funds combined rose from £89 million in 1965 to £633 million in 1973. The property slump, and the era of lower rent rises which followed, saw paradoxically a sharp increase in property investment by the financial institutions. But from acting simply as lenders they were moving to a more pro-active role, initiating development themselves, becoming more active as freeholders of sites leased for development and taking ownership shares in the many property companies who were having difficulty repaying their loans. For example Marler Estates sold its Knightsbridge estate to the BP Pension Fund for £45 million – a 'knock-down' price. Many such bargains were available since the property share index in November 1974 had fallen to less than one-quarter of its value a year earlier.

It is important to remember that investment in property schemes forms only a small part of total institutional investment. Even in the peak year of 1973, property investment accounted for only 18.5 per cent of insurance company investment and 26.3 per cent of that by pension funds (Rose, 1985,

170). More typically through the 1970s and early 1980s the proportion was in the range 10–15 per cent. Since then striking changes have occurred in the source of property finance. Net institutional investment in property, which had peaked at nearly £2 billion, fell to £0.5 billion in 1987 and has since fallen further until it represents perhaps only 3–5 per cent of an institution's new investment. There are a number of reasons for this. The fall in the rate of inflation has reduced the attraction of property investment, which has often been regarded as a hedge against inflation. The return obtainable on equities and gilts rose faster than that obtainable on property investment through the 1980s. And the rapid spate of property investment through the later 1970s left many funds feeling that they were over-invested in what has often proved to be a very illiquid asset. As we have seen, the latest mid- and late 1980s upsurge in the rate of development has been financed mainly by banks acting primarily as lenders rather than institutions taking a proactive role via the ownership of property shares or particular schemes.

The current result of these decades of investment is that the total ownership of property assets is relatively concentrated. There are about a hundred investing institutions with property assets in excess of £100 million, and these control, between them, about 90 per cent of institutional property assets – in fact the largest ten own nearly one-half of the total (Healey and Nabarro, 1990). The insurance companies have a 60 per cent share, the pension funds 35 per cent and Property Unit Trusts about 5 per cent. Smaller 'historic' property investors such as the Crown, the Church and the universities, own a small residual share.

Construction

Firms in the construction sector carry out the actual building in accordance with a development contract. Work is put out to tender, either to a selected list of contractors or by means of an open invitation to tender. The contract may be individually negotiated, or it may form part of a 'serial' sequence of contracts when a series of similar jobs needs to be carried out and one contractor has successfully completed the first in the series. Contractors may also be involved to varying degrees in the design aspects of the scheme, depending on the nature of the contract with the developer. For example the JCT (Joint Contracts Tribunal) type of contract provides for the main contractor to build to the developer's designs and the 'Design and Built'-type contract, as the name implies, is for the design as well as the construction of the scheme. The latter type of contract is usually used for simpler forms of development. The various possibilities are well discussed in existing literature (see, for example, Balchin, Kieve and Bull, 1988, chapter 10; and Cadman and Austin-Crowe, 1991, chapter 6).

The main issues that need to be specifically agreed upon between the developer and the contractor include the cost, the rate at which payments are made to the contractor, the means by which cost inflation is built in, the monitoring of the construction so that it conforms to the agreed specification, and the penalties to be imposed for late completion or some other failure to fulfil the contract. Since the early 1980s the specialist task of 'management contracting' has become more important. The management contractor, who is involved at an early stage with the developer's design team, offers a package deal by breaking down the total task into a number of trade-specific contracts. He/She is then answerable to the developer for their successful completion. If the project is a large one, a project manager may be appointed, either for a proportion of the total development cost (perhaps 2 per cent) or on some basis that provides an incentive for cost minimisation. Such a person may be appointed at a very early stage and may assist the developer in the selection of the design team. He/She will then probably be responsible for site supervision during the construction period and up to the handover stage.

The construction industry is a sizeable element in the British economy, employing, by the late 1980s, over 900,000 people (the total had been about 1,468,000 in 1970). It is extremely sensitive to central and local government policy because a substantial proportion of its work still comes from public bodies (see Chapter 9). Apart from this, the industry suffers during periods of high interest rates, as in the mid- to late 1980s, because so much development is loan-financed, and expensive money brings about sharp cost increases. Not only that, but house purchasers and users of commercial space have less available to spend on buying or renting property if high interest rates are hitting their finances in other ways. The industry also experiences business downturns when public spending is reduced, especially when capital programmes are postponed or cut. Despite considerable urging from the industry and other sources there has been little use in recent decades of the Keynesian notion of using investment in construction as a counter-cyclical device, perhaps because the 'lag' factor built into property schemes is both long and unpredictable, and partly because, as Chapter 8 will show, this is not what neo-liberal governments believe in.

THE TYPE AND DISTRIBUTION
OF PROPERTY ASSETS

The distribution of institutionally owned investment properties at any given point in time reflects a long history of urban development. It also explains much about the dynamic processes that shape the built environment, since every single investment was made only after a careful evaluation, certainly by the promoter and perhaps by the investor, not only of the effective demand for a building of that particular type but also of its prospects for success in that location. Arguably these interests, in the course of making such decisions, do as much as any other to condition the evolution of the built environment. It is they, in other words, who have much of the power.

At the end of 1988 offices accounted for 53 per cent of total property assets held by financial institutions, retail developments for 36 per cent, and industrial property for less than 10 per cent (Healey and Nabarro, 1990). This distribution followed a period of disinvestment in offices and an increase of holdings in shops. The office properties are very much concentrated in central business-district areas, and over two-thirds of them are freestanding blocks of fewer than ten storeys. The high-rise prestige office block is very much the exception. Over half of institutionally owned retail property assets takes the form of parts of, or the whole of, shopping centres; individual shop units are much less important. There is also very little institutional investment in supermarkets and department stores. These are often developed and financed by the retail chains themselves. Industrial property investment is mostly in warehouses with conventional manufacturing premises representing only a quarter or so of the total value in this sector. In terms of geographical distribution, London and the rest of the south-east dominate the investment-property scene, with the City and the West End alone accounting for 64 per cent of national investment in office development in 1987.

CURRENT TRENDS IN RENT LEVELS

As we have seen, rents are the main source of the $M' - M$ increment which forms the motivation for all activities in the investment property field. The anticipated size of the future rental flow is the starting-point for the financial appraisal of any scheme (see the Case Study in this chapter) and in most cases plays a large part in setting the value of the site on which the development is to take place. It is therefore vital to take account of past trends in rentals and to seek to make forecasts that are as accurate as possible about their future behaviour. Information on rent trends is collected and made available by a number of firms, including Debenham Tewson Research, Jones Lang Wootton, Richard Ellis, and Hillier Parker Research. Some smaller property analysts tend to specialise in researching 'niche markets' such as high-street shops.

Rents vary enormously not only by type of use but crucially by location. Naturally the characteristics of a preferred location are different for each of the three main types of investment property. The attributes of a good office location for, say, financial services would include centrality in a large financial centre, good public-transport accessibility, a 'prestige' address and good nearby restaurants and other lunchtime facilities for staff. For most retail developments, although not retail parks, the key factor is the number of pedestrians passing the site. This can vary considerably over distances as small as a hundred yards or less, depending on the detailed configuration of the street pattern and the location of facilities such as railway and bus stations. For industrial property accessibility for the movement in and out of goods is often crucial. For all three categories, locations are divided into 'prime' and 'secondary', and trends can be distinguished in, say, 'prime' offices or 'secondary industrials'. The assessment of future rent trends, therefore, has to take account not only of projected changes in the overall level of demand for each category of space, but also of any changes that are becoming evident in locational preferences.

The ICHP rent indices from Hillier Parker for selected regions and years give a good indication of rent movements for the past three decades.

Table 4.4 Rent indices for various uses and areas 1965–92

	1965	May 1977	May 1990	May 1992
All shops	29	100	604	572
Central London	14	100	486	443
Suburban London	38	100	563	519
South east	27	100	666	610
East Anglia	32	100	735	725
Scotland	21	100	462	480
All offices	28	100	489	336
Central London	28	100	645	369
Suburban London	20	100	392	312
South-east	27	100	416	345
East Anglia	26	100	421	370
Scotland	22	100	385	464
All industrial	28	100	397	359
London	32	100	461	404
South-east	31	100	494	434
East Anglia	35	100	463	407
Scotland	17	100	179	223

Source: Hillier Parker/Investors Chronicle, *ICHP Property Market Indicators*, February 1990

These data, which are of indexed trends not absolute rent levels, reveal a great deal about the changing fortunes of different sectors and regions. Shop rents have risen faster than any others over the past fifteen years as rising real incomes for the majority and a massive extension of credit have been translated into increased consumer spending. Office rents have risen less sharply but with greater regional variability. For example the index value in May 1990 was 235 in the north and 872 in the West End of London. Since the end of the late 1980s boom, virtually all indices have fallen in nearly all regions, with the exception of Scotland, where all three indices have continued to rise. In the early 1990s recession the fall-back of office rents has been especially marked in London and the south-east, with exceptionally

Plate 4.9 Signs of the late 1980s/early 1990s business recession on a Newhaven retail park. This development was on land owned freehold by Lewes District Council and made available on a long lease to a development company who then attracted the retail tenants. This particular store sold carpets and had a good range at keen prices, but a competing retail use in the shape of a supermarket is taking over the site, and much of the space will be required for parking. It may well be that in a recession food sales hold up better than carpet sales. Photograph by Peter Ambrose.

severe fall in all parts of central London (see Debenham Tewson Research 1990c, 1992a, 1992b, 1992c; and Harris, 1991).

As at July 1993 the trends shown in Tables 4.3 and 4.4 were continuing. The London West End rent level had fallen to £40 and that of the City to £30 (Richard Ellis, 1993). City rents are now less than one-quarter of those in the Tokyo Central Business District and, perhaps more significantly, below those of rival European financial centres such as Paris, Berlin and Frankfurt. They are only marginally above those in Lisbon and Milan. These figures relate, of course, to properties that are let. A very sizeable percentage of London property remains unlet and is earning nothing at all. The implications of these very sharp falls in rent revenues in recent years for the financial institutions and other interests who have invested in commercial property can be gauged from the Case Study in this chapter. Banks, pension funds and life-assurance companies who invested the money of depositors and contributors in major property schemes on a rent expectation of, say, £70 per square foot are clearly sustaining major losses if the rents actually achievable range from £30 to zero. The key political questions are how this situation arose and how the losses are being spread.

Data on rent levels in the major urban and suburban centres of Britain are available from the Jones Lang Wootton *50 Centres* series, which is produced twice yearly. In the office sector 'rents achievable' as at September 1992 ranged from £25 in Hammersmith to £10 in Wembley and Bournemouth (all figures are in pounds per square foot per year). London Docklands is not separately identified; presumably many investors would currently be relieved if even £10 were achievable there. In terms of rental growth from 1979 to 1992 the best centres were Norwich, Plymouth and Chelmsford. In the industrial sector the top rent was £10 in Maidenhead and the lowest £2.35 in Liverpool. The five fastest-growing areas made a neat pattern along the M4 – Maidenhead, Uxbridge, Hounslow, Windsor and Slough. Retail rents (for 'Zone A' – prime locations) ranged from £180 in Chester and Manchester to £70 in Middlesbrough. Rents had grown fastest over the past thirteen years in Manchester, Luton, Hull, Chester and Swindon.

It must be remembered that data such as these are highly generalised. Each urban centre is a specific market with specific characteristics and problems. Brighton, for example, a town on the south coast with a 'catchment area' of approximately 400,000 people, has a stock of vacant offices totalling about 430,000 (CSW, 1993). There has been a recent sharp decline in the rate of take-up, and very few new lettings of any size have occurred since September 1991, when some space was taken at £23.50 per square foot. Prime modern office space is now on offer at £10–15 and refurbished space slightly off-centre at £7 or so. There are few enquiries for either, and to compound the problem a new development of 250,000 square feet adjacent to the station is coming on to the market in late 1993. The market for retail space is similarly depressed. Estate agents claim that this is partly due to the town's restrictive, if environmentally friendly, parking policy. The main shopping centre, Churchill Square, which was developed by Standard Life over twenty years ago, is now outdated and has lost major retailers such as Tesco and Sainsbury's to more peripheral locations. Brighton Council as freeholders and planning authority, and Standard Life as leaseholders, are currently discussing the possibilities and finances of upgrading the centre. Meanwhile the fall in interest rates means that more retailers can consider buying shop units, thus depressing the rental market still further. This brief review shows how a range of factors – economic, political, environmental, behavioural and technological – combine in complex ways to affect markets in investment property.

CASE STUDY – THE CITYGATE DEVELOPMENT

This is a fictitious development combining the details of several real cases. It sets out the financial assumptions of a proposed mixed-use speculative scheme for renting, and the calculations by which a bid to acquire the site is arrived at. The gross area of the building will be 320,000 square feet and the lettable space 258,000 square feet. Planning consent for these figures, and the uses proposed, is expected. The site is towards the fringe of the City of London, and because of the sensitive nature of the local townscape the design will need to be agreed with various conservation bodies. The present freehold owner of the site is an old-established City company. The bank finance for the period necessary to obtain planning consent and tenders (eighteen months), to carry out the construction (twenty-four months) and to let the building (twelve months) is expected to be available at 10 per cent. When completed and let, the scheme will be sold to a longer-term investor, probably a pension fund or life-assurance company. This institution, it is assumed, will be looking for an annual return of 5.75 per cent on the price it pays for the scheme. This is known as the 'yield' – the annual rate of return sought on an investment with this degree of risk. Building costs are estimated at £190 per square foot. The developer is seeking a return for risk and reward of 20 per cent on the money spent on the construction and related costs. The various professionals, contractors and sub-contractors will include a profit element in their tenders for the contracts.

Building costs calculation

320,000 square feet construction @ £190 psf	60,800,000
Professional fees, architects, etc. @ 12%	7,296,000
	68,096,000
Funding costs @ 10% (averaged for half the construction period)	6,809,600
	74,905,600
Finance @ 10% during 12-month letting period	7,490,560
Letting fees to agents @ 5% of total expected rents	612,750
Legal and other fees	190,620
	83,199,530
Costs to sell the scheme @ 1.5% of gross value	3,109,040
Total construction, etc., cost	£86,308,570

Site-bid valuation

Annual rents expected (the space is offices unless otherwise stated):

Storage space	2000 sq. ft. @ £15 psf	30,000
Lower ground floor	23,000 sq. ft. @ £25 psf	575,000
Ground floor	9000 sq. ft. @ £50 psf	450,000
Upper floors	200,000 sq. ft. @ £50 psf	10,000,000
Retail space	24,000 sq. ft. @ £50 psf	1,200,000
Total annual rental income projected		12,255,000
Capital value of scheme @ 5.75% yield		213,130,435
Less purchaser's costs @ 2.75%		5,861,087
Gross value of development when let		207,269,348

Less:

Developer's return @ 20% of total construction, etc., cost	17,261,714	
Total construction, etc., cost	86,308,570	103,570,284
Available for site purchase and purchase finance		103,699,064
Less site purchase finance for 54 months @ 10%		32,182,466
		71,516,598
Less site purchase costs @ 2.75%		1,914,070
Site value at date of calculation		£69,602,528

These figures derive from calculations first made in the market conditions of early 1991. What happens to the capital value of the scheme, the site value and other aspects of the finances if construction costs rise by 20 per cent, or the yield required by a potential purchaser of the completed scheme increases to 8 per cent, or the cost of short-term finance increases to 15 per cent, or office rent levels fall by 40 per cent, or only half the space can be let – or all five happen together? Such changes in market conditions during the development period have been commonplace in recent decades. What changes in conditions would it take to make the value of the completed scheme less than the construction costs? If this happens, who is likely to bear the loss? And what action might the losers take to recoup the loss?

5

PROFIT-SEEKING DEVELOPMENT – FOR SALE

The twentieth century has seen an immense growth in this type of development in Britain. Home ownership, restricted to the very rich until the early years of the century, had become by the 1990s the tenure form of about 70 per cent of the population. The financial, legal and other issues surrounding owner occupancy are so embedded in the political, social and economic life of this country that it is difficult to deal with the provision processes without taking some account of the historical and ideological context in which this growth in ownership has occurred. The literature concerned with owner occupancy is enormous and cross-disciplinary and cannot even be summarised here. It includes technical discussions of construction techniques, historical accounts of the growth of the lending institutions, financial analyses of the effect of changes in house-purchase credit on the broader economy, discussions concerning the effects of ownership on political beliefs and voting patterns, and even disputes about whether or not there is an innate desire to own property in order to meet some psychological desire for personal security.

The issue has acquired some of these dimensions partly because with the rise in property ownership housing has taken on several new meanings and functions. The simple act of paying money periodically to acquire shelter, as one does when renting, has had overlaid upon it investment and display motivations. The need for shelter has become interlinked with the desire to acquire and protect a personal capital asset and to display wealth. In fact housing must be one of the few products where high and rising prices are generally regarded as beneficial to users rather than the reverse. Judged in terms of some expected capital gain by existing owners this attitude is understandable. Judged from the viewpoint of those who cannot afford to buy, or in terms of the proportion of one's lifetime earnings that has to be expended to make the purchase, it is not.

Most of those who advise new entrants to the housing market, including professionals in the legal, estate-agency and financing fields, stress the desirability not only of ownership, but of making as large as possible an investment in purchasing one's home. They have good reasons for doing this because their fees and interest income are directly related to property values. In purely investment terms this may have been good advice so long as house values were rising faster than general inflation, which was broadly true until the late 1980s. In the post-1988 era of falling or at best stable house values many people have been obliged to think again (see, for example, Took and Ford, 1987). But house-purchase decisions are not simply about investment; they carry implications for one's lifestyle, pattern of expenditure and capacity for industrial action during the twenty-five years or more of indebtedness that normally follow the purchase.

THE PRO-OWNERSHIP IDEOLOGY IN BRITAIN

The rapid expansion of owner occupancy in twentieth-century Britain has taken place in an increasingly pervasive pro-ownership climate. Thus the act of purchasing one's home has become far from ideologically neutral. It denotes 'proper behaviour', a personal step to help bring about a 'property-owning democracy'; it reflects 'thrift' and it is evidence of 'standing on one's own two feet'. Those who depend on the main alternative housing arrangement, renting from a local authority, have come to be seen as somehow less than proper citizens fecklessly benefiting from disproportionate amounts of state subsidy. They have been widely depicted as 'unable to compete in the housing market' or as 'social cases' in a phrase once used by Mrs Thatcher to characterise, or more accurately ghetto-ise, council tenants.

The housing system as a whole is frequently portrayed as a 'ladder' up which one climbs having got on to the first step as a 'first-time buyer'. Almost anyone, it is argued, can save the necessary deposit, take out a loan, buy in at the bottom of the ladder and subsequently 'trade up', leaving the cheaper properties for those who subsequently join the ladder. Thus ownership is portrayed simultaneously as a reward for thrift, as a personal investment, as a means of saving public money, as a way of achieving social and locational mobility and as the proper and moral way to deal with one's personal housing needs. It is worth noting that the association of this set of virtues with a particular form of housing tenure is virtually unknown elsewhere in Europe. Well-heeled Swiss or Germans (owner occupancy rate 30 per cent and 37 per cent respectively – see *Roof*, January/February 1991) are quite happy to rent or buy into a co-operative and by 1993 the slogan 'Rent yourself free' was gaining political mileage in prosperous Sweden (owner occupancy 39 per cent). There is also considerable popular pressure in Spain (owner occupancy 77 per cent) for an enlargement of the rented sector so as to bring about more housing choice.

This pro-ownership ideology, heavily projected in the 1980s in Thatcherite

Britain, is now revealed as a political construct underpinned by some very identifiable interests and some major but unstated assumptions. The assumptions are that nearly everyone is in secure and lifelong employment with a rising real income, that the household that took out the loan will not break up during the repayment period, that general financial conditions, and especially the rate of interest, will not materially change, that the market in sales and purchases works efficiently, that a house as an asset will rise in value faster than any alternative investment and that tenants of public housing are subsidised whereas those who purchase are not.

The reality, during the same period, has undermined or falsified every single one of these assumptions. There has been mass unemployment, falling real incomes for many, increased family break-up, sharp changes in interest rates, a stagnant housing market and falling property values. Finally the notion that council tenants receive more subsidy per capita than owner occupiers has been revealed as a reversal of the truth. The pro-ownership ideology is therefore increasingly being called into question. It will be argued in a later chapter that British housing analysis and policy have been bedevilled by an obsession with tenure distribution and that this has detracted from more sensible activities such as evaluating the cost-effectiveness of alternative provision processes and subsidy programmes. As we have seen, home ownership has not been fetishised in this way in other Western European nations. So why is Britain different? Where did the obsessive concern with owner occupancy originate, why has it been so resilient and what have been its social and political effects?

Some of the standard works on housing history (for example Burnett, 1991) are unforthcoming on these issues. Others (see Mayes, 1979, chapter 3; or Saunders, 1990, chapter 4) discuss them at length. Various aspects of the pro-ownership ideology have obvious roots in the building-society movement. Building societies have dominated the house-purchase lending market, and the movement has, increasingly, exerted political influence at high levels. It is these institutions too whose growth depends crucially on the growth of home ownership and prices, since, until very recently, lending for this purpose was overwhelmingly the main means by which they expanded (Boddy, 1980). There appears to be no parallel case in housing-finance systems elsewhere of this combination of market dominance and institutional specialisation. If the British building-society movement is unusual in an international context, it is reasonable to conclude that it is at least partly the influence of this movement that helps to account for some of the oddness in British housing history and especially the obsession with ownership as the only 'proper' tenure.

THE BEGINNINGS OF MASS HOME OWNERSHIP

Before 1914, when nearly all housing was produced for renting, the rate of output of new housing tended to be related to the apparent prospects for a profitable rate of return, in other words to trends in rent levels. Since the mid-nineteenth century these had risen irregularly but on the whole faster than other prices (Lewis, 1965). In the periods of sharp upturn, which were likely to be periods of real increases in incomes, building activity tended to increase – although the level of activity was very different area by area (Dyos, 1961). This reasonably direct relationship between incomes, capacity to pay rent and housing output has become more complicated with the growth of purchasing based on long-term credit as opposed to renting. The new intervening variable is the volume and cost of the flow of house-purchase credit. The volume may increase relatively suddenly as a result of new levels of fund availability, new lending practices, or some other factor even though no change has occurred in real income levels. Thus the effective demand for the product might increase or decrease to some extent independently of levels of earnings. This underlines the crucial importance of the lending institutions to the understanding of the history of the speculative housebuilding industry.

The origins, gradual transformation, growth and aims of the building-society movement have been exhaustively discussed (see, for example, Boddy, 1980; Merrett with Gray, 1982; Ball, 1983; Barnes, 1984; Boleat, 1985; Forrest, Murie and Williams, 1990; and Saunders, 1990). Various of these commentators have focused on the relatively sudden expansion of spec-ulative housebuilding for owner occupancy which occurred in the years following the Great War of 1914–18. It is instructive to consider the political and social conditions prevalent across Europe in the early 1920s. In 1917 the Romanov dynasty was ended in Russia and replaced by the socialist Soviet Union. Revolutionary fervour was at work in the Balkans and in Germany. No one could be sure whether or not the post-war radicalism would spread as far as Britain. During the war a Royal Commission was set up to enquire into the causes of industrial unrest. It had reported that poor housing conditions in several industrial areas were near the root of the problem. The Increase of Rent and Mortgage Interest (War Restrictions) Act, passed in 1915, following serious agitation on Clydeside and elsewhere, had intro-duced for the first time the principle of legislatively controlled private-sector housing rents – a clear negative signal to those considering the investment of funds in rented housing. From their point of view the prospects were worsened by two further acts, in 1917 and 1918, giving greater protection against the wave of evictions, news of which was having a disturbing effect on the morale of the troops at the front. So once victory had been achieved how were housing conditions, clearly a source of political destabilisation, to be improved?

81

The post-war election had been fought partly on the issue of 'Homes Fit for Heroes'. King George V himself, in a speech in April 1919, made the point: 'If unrest is to be converted into contentment the provision of good houses may prove one of the most potent agents in that conversion' (Bellman, 1928, 29). The post-war coalition government introduced the Housing, Town Planning, etc., Act of 1919, also known as the Addison Scheme. This offered local authorities generous and relatively open-ended subsidies to carry out slum clearance and public housing schemes (the issue will be dealt with more fully in the next chapter). But the cost to the Treasury was soon judged to be excessive and the scheme was ended in 1921. The way was open for those who spoke for the building-society movement and the growth of home ownership. One such was Harold Bellman, who wrote a short book entitled *The Silent Revolution* – presumably to distinguish this process of change from the noisier revolutions that had occurred in Russia and elsewhere:

> The Building Society not only assiduously preached the doctrine of self-help, but in the minds of men emerging from the travail of unrest it inspired a new conception of citizenship. By assisting men to the ownership of houses it gave countless thousands a stake in the country and enlisted their aid in the struggle against those forces of unrest and unreason which seek to uproot and destroy and to substitute extreme theories which, however politically desirable, are economically impossible.
>
> (Bellman, 1928, 34)

Mrs Thatcher could hardly have put it better although she would not have conceded any political desirability to the 'extreme theories'. Similarly Samuel Smiles: 'The accumulation of property weans men from revolutionary notions' (ibid.) Similarly Neville Chamberlain, when minister responsible for housing: 'every spadeful of manure dug in, every fruit tree planted converted a potential revolutionary into a contented citizen' (Feiling, 1946, 86).

It was not only the political conditions that were right for a rapid expansion of owner occupancy. The years leading up to 1914 had seen declining levels of housing output and very little was built during the War. Yet the conflict had brought a quickening of economic activity and placed a considerable reservoir of purchasing power into the hands of those who had benefited. Many of the demobilised fighting forces, too, had their gratuities on leaving the colours. The building societies achieved great success in attracting savings and they began to expand their lending rapidly. Whereas in 1913 £9.1 million had been loaned for house purchase, the figure rose to £19.7 million in 1921, £32.0 million in 1923 and £51.1 million in 1926 (Bellman, 1928, 33). The Housing Act, 1923 brought the local authorities in as alternative lenders to purchasers, and as lenders and providers of a subsidy and guarantees to builders. But whereas by the beginning of 1927 they had

SIGNS OF SPRING CLEANING.

Plate 5.1 A cartoon, origin unknown but dated around 1920, summing up the political significance accorded to investment in property ownership. As the text makes clear, politicians such as Lloyd George and Neville Chamberlain, to say nothing of King George V himself, repeatedly made clear that they regarded an expansion of owner occupancy as a factor that would reduce the risk of political instability in Britain. A spokesperson for the building-society movement even called it a 'bulwark against bolshevism' (Bellman, 1928). Here the 'cleaning materials' of savings, deposits and bonds are about to eradicate a variety of social ills. Now, with owner occupancy at around 70 per cent some, but not all, of them have been washed away.

been sanctioned to lend out about £33 million for house purchase, the building societies had by that time lent out £235 million (ibid., 35).

There were other factors underpinning the early growth of owner occupancy (see especially Merrett with Gray, 1982). These included rises in real disposable income per head in the inter-war years for those in work and an increase in the population in the household-forming age groups. Building society loans outstanding rose from £120 million in 1924 to £636 million in 1937 (Bowley, 1945). The availability of loan funds was dependent on the societies' attractiveness to savers which, in turn, reflected the combination of liquidity, security and high deposit rates they were able to offer. They also looked, and indeed were, 'local' and this factor helped to build up saver loyalty. Borrowing from building societies for house purchase was made more attractive by various 'special arrangements' concerning taxation which reduced both the societies' tax liability on interest received from borrowers and benefited the borrower by allowing the interest on loans to be tax deductible (see Boleat, 1985, chapter 13). In addition there was no Capital Gains Tax levied on gains from the purchase and sale of the house one lived in. During this period, too, lending practices changed. The societies began to offer loans based on a higher proportion of the valuation and repayable over a twenty-five- rather than a fifteen-year period. Finally, in the early 1930s the mortgage-loan interest rate fell from 6 per cent to 4.5 per cent.

The effect of these changes in combination was to allow a larger capital sum to be lent in relation to any given weekly or monthly income. This enabled those further down the income scale to come forward as potential buyers while still allowing builders to achieve stable or even rising prices for their product. In terms of production costs, potential housing land on the fringes of built-up areas was often available relatively cheaply since farming was a depressed industry for much of the inter-war period. But simultaneously rising rates of car ownership, and public transport improvements, were making these areas more accessible (see Jackson, 1973; and Ambrose, 1986, 16–22). In addition, and reflecting the high rates of unemployment, the wages in the building industry fell in real terms. As a final factor, investment in the private rented sector fell as many new middle-income households turned instead to owner occupancy and the prospect of a rising equity stake in their home. All the factors were working in the same direction and they led to a national home-ownership rate of about 32 per cent of the total stock by 1938 – although the figure varied considerably by town from 14.3 per cent in Nottingham to 68.5 per cent in Plymouth (Forrest, Murie and Williams, 1990, Table 3.3). In all there had been a remarkable shift in tenure patterns since 1919. It is reasonable to argue that the broad pattern of events was as much dependent on credit availability as on anything else. It thus turned crucially on the growth and behaviour of the main lending institutions – as has frequently been the case since.

It was in these inter-war years that a major housebuilding industry first emerged. The early growth has been well documented (see especially Jackson, 1973; Ball, 1983; and Burnett, 1986, chapter 9). Some firms such as Wimpey had been founded before the Great War and expanded rapidly in the post-war boom. One such, Wates, built up to volume-builder status by specialising on particular locations such as those within walking distance of a Southern Electric railway station in south London. Others such as New Ideal Homesteads were set up in the 1920s, while firms such as Laing and Costain moved from elsewhere in the country to benefit from the demand in London and the south-east. Others such as George Ferguson, who built 1300

Plate 5.2 Terraced houses built speculatively for owner occupation in the 1930s. This estate is in south-east London, within a few minutes' walk of a former Southern Electric railway station. Typically built for sale to 'bus-drivers, postmen and policemen' – or any buyers on a fairly secure income of £3–4 per week – they were purchased for maybe £300–350 with the aid of a loan from the rapidly expanding building-society movement. Their original uniformity has now been modified by the home-improvement industry, and attic rooms, replacement windows and stone-cladding abound. Photograph by Keith Hunt.

Plate 5.3 The fully transformed 1930s spec.-built terraced house. Reading from the top down, the satellite dish, attic room, leaded-light windows, burglar alarm, new front door, late-registration Mercedes, 'landscaped' garden, metal gates and paved forecourt announce that this house has come up in the world and should really be in the 'stockbroker belt'. Photograph by Keith Hunt.

85

Plate 5.4 This illustration shows some of the possibilities for adapting the standard 1930s semi. These houses were built in Patcham on the outskirts of Brighton on farmland bought cheaply by the developer in 1931. They sold originally for £550, with 10 per cent down and mortgage repayments of 16s (80p) per week over twenty years. Over 1300 houses were sold within a three-year period. They have proved to be a flexible form of housing. Some have had extra bedrooms added in the roofspace. The original finishes have been transformed. Some have been given a flintstone finish, some are now stone-clad, and the furthest one has decorative long and short work at the corners. Nearly all have replacement windows, and none of the porches or garden walls is original. Photograph by Peter Ambrose.

houses at Patcham on the northern fringes of Brighton in the early 1930s (Ambrose, 1986, 18–19), started as tradesmen and gradually built up their own construction businesses.

All benefited from the expansion of purchase credit which enabled them, to quote one Wates director, to build for the bus driver, the postman and the policeman earning perhaps £3–4 per week. By the early 1930s a two-bedroom terraced house might be priced at £295 and a three-bedroom semi at £395. Even in the late 1930s builders such as Martin and Saunders of Pevensey near Eastbourne were selling seaside bungalows to members of the inter-war nouveau riche for £270. The cheaper properties might typically be bought for a £5 deposit and repayment of a twenty-one-year loan at the rate of 7s 6d per week. In these conditions of rapidly expanding demand output rose from 118,000 units in 1928 to a peak of 293,000 in 1936. Production then began to fall as building costs, long depressed in the worst of the recession, began to rise and the market at the resultant higher price levels began to show signs of saturation. The activities of the industry were finally halted by the outbreak of war.

THE POST-1950 EXPANSION
IN OWNERSHIP

A second burst of expansion in the market for owner-occupied houses took place from the early 1950s onwards. Under the 1945–51 Labour governments, the emphasis had been on public-sector-led housing reconstruction. The Housing (Financial and Miscellaneous Provisions) Act, 1946 assisted local housing authority capital programmes by increasing the level of subsidies and rate-fund contributions and the 1949 Housing Act signalled intentions of a 'general provision philosophy' by removing the statutory requirement that public housing should be for only 'the working classes'. As a result in 1945–51 over 80 per cent of new housing was produced by local authorities. But already by 1948, under the terms of the Marshall Aid Plan providing US help to reconstruct Britain's economy, public-sector spending programmes were under threat (Cooke, 1957).

In 1951 the incoming Conservative government set a high output target of 300,000 houses per year, to be met by public and private sectors in combination. But between 1954 and 1961 the public-sector output fell from nearly 224,000 to fewer than 106,000. Meanwhile the output of the private housebuilding industry expanded steadily to reach annual peaks of 220,000 units or more in the mid-1960s. This led to the rate of owner occupancy reaching 50 per cent in 1970. This milestone was very significant politically. It made it very unlikely that a General Election could subsequently be won on any programme which appeared, even remotely, to threaten the interests of owner occupiers. It is a tenable argument that this electoral circumstance has, ever since, been one of the main impediments to the open political discussion and implementation of more cost-effective and equitable housing policies.

So in what context did this further drive to majority home ownership come about? As always a number of factors contributed. Building materials were freed from control in 1953, and the same year's Town and Country Planning Act abolished taxation of the gain in value as land passed from agricultural to building use. This encouraged both landowners to sell and builders to build up landbanks for future requirements. The number of building society advances, which had rarely topped 300,000 per year in the 1940s, had reached 387,000 by 1960 (average value £1447) and 624,000 (average value £3131) by 1970 (*Annual Reports* of the Chief Registrar of Friendly Societies, HMSO, London). This increase in both numbers and average value reflected the rising real incomes and sound employment prospects of the 1960s. The trend continued and by 1978 the number was 1,184,000 (average value £7439). Since virtually all houses are bought with the help of a loan, the figures illustrate the rate of growth in effective demand for the product of the housebuilding industry.

Meanwhile both major parties were encouraging owner occupation. Thus the Conservatives:

One object of future housing policy will be to continue to promote, by all possible means, the building of new houses for owner-occupation. Of all forms of saving, this is one of the best. Of all forms of ownership this is one of the most satisfying to the individual and the most beneficial to the nation.

(*Houses: The Next Step*, Cmd 8996, 1953)

Britain's subsequent economic performance may lead one to doubt whether the use of so much private-sector saving for this purpose has indeed been beneficial to the economy. But similarly Labour, soon after their election to office in 1964: 'The expansion of building for owner-occupation ... is normal; it reflects a long-term social advance which should gradually pervade every region' (White Paper, *The Housing Programme 1965 to 1970*). By this time the percentage of electors who were home owners was over 46 per cent and rising fast. Any policy line proposing public-sector rented housing as the favoured long-term tenure form would already have been politically hazardous. By the mid-1970s the Secretary General of the Building Societies Association was quite explicit about the political implications of home ownership:

The point where more than half the houses in the country had become owner-occupied was a significant milestone because even a small stake in the country does affect political attitudes. The greater the proportion of owner-occupiers the less likely were extreme measures to prevail.

(Griggs; quoted in Ambrose, 1976)

There must presumably be other reasons apart from this one why Labour has not won a General Election since.

The structure and operations of the speculative housebuilding industry in the post-1950 period have been discussed elsewhere (see, for example, Ball, 1983 and 1988; Centre for Urban and Regional Studies, 1981). It is a functionally separate part of the construction industry with a number of specific characteristics. Although for much of the period perhaps half the annual output of new houses was produced by a small number of 'volume builders', the industry has been dominated numerically by small firms. The *Private Contractors' Construction Census 1978* showed that 40 per cent of new private-housing work was being done by firms employing fewer than twenty-five people – a degree of fragmentation unknown in most major industries.

Most of the labour engaged in housebuilding is not permanently employed by builders, but subcontracted for particular jobs or schemes. This arrangement reflects partly the extreme fluctuations in levels of activity, since firms employing a large permanent workforce experience over-capacity during downturns. It also reflects the degree to which housebuilding involves specialist trades such as plumbers, roofers, plasterers, and so on. At the lowest end of the size scale, considerable output during upturns is produced by very small or one-man building companies. These may revert to their original trade, or set up business in some sector of the materials-supply industry, when the demand for houses falls. One consequence of this highly fragmented pattern of ownership and the unreliability of demand has been that the industry has invested less in research, modern plant and new technology than industries of comparable size such as the vehicles industry.

THE 1980s – THE MORTGAGE BOOM AND ITS EFFECTS

By the time the first Thatcher administration was elected in 1979, 55.3 per cent of homes were owner-occupied, 31.5 per cent rented from public authorities and 13.2 per cent were other tenure forms, mostly privately rented (see Forrest, Murie and Williams, 1990, Table 3.1). The 1979 Conservative Election Manifesto promised that public tenants would have the right to buy their homes at discounted prices and that 100 per cent mortgages would be available at lower interest rates. The beneficial effects of ownership on personal responsibility, independence and pride, the better care of property and the general stability of society were expounded from the beginning (Conservative Central Office, 1979). The Conservatives' top housing priority, therefore, was to increase the rate of home ownership by bringing about a change in the tenure balance. It was not, it might be noted, to solve the existing problems of homelessness, improve the widespread bad conditions, increase housing output, make more cost-effective use of public subsidy or even foster a more stable commercial environment for house-builders – many of whom began the 1980s as good friends of the government but ended the decade as sceptics or outright enemies.

During this decade policies to increase the ownership percentage were vigorously and consistently pursued on several fronts. For a number of

Table 5.1 Balances outstanding for house-purchase loans

Year end	Amount (£bn)	% building society
1980	52.4	81.5
1982	76.4	74.9
1984	107.9	76.6
1986	150.7	77.4
1988	226.7	68.4
1990	294.4	60.0
1992 Q1	325.3	61.7

Source: Council of Mortgage Lenders, *Housing Finance*, no. 15

commentators (for example Hamnett, 1987, Merrett, 1991) these policies had more to do with ideology than with economics. The Housing Act, 1980 introduced the right for council tenants to buy their homes, a right strengthened by subsequent acts in 1984 and 1986 (see Malpass and Murie, 1987, Table 5.2). As a result over 1.4 million council and new town corporation dwellings have been sold. At the same time, public-sector capital investment in housing construction was savagely cut and local authorities were empowered to use only a small proportion of their receipts from sales to finance building programmes. These measures constituted a partial closing-down of the alternatives to owner occupancy. Further policy measures to expand owner occupancy centred on increasing the availability of house-purchase credit. Unfortunately the manifesto promise of lower mortgage-interest rates fell foul of other aspects of macro-economic policy. Under the prevailing monetarist orthodoxies, the government's management of the economy was for a period confined largely to regulating the supply and cost of money. This meant that the interest rate had to be adjusted up and down to serve a number of purposes, for example the defence of sterling, which had nothing to do with the housing sector. Thus the average rate of interest charged on building society mortgages varied from 10.97 per cent to 15.12 per cent at various times between 1980 and 1990 (it had been 9.55 per cent in 1978).

But if promises about the cost of house-purchase credit could not be easily delivered, the supply of credit was sharply increased during the 1980s. The deregulation of the banking system in 1980 enabled the building societies and banks to compete more freely in each other's traditional markets (Boddy, 1989, Ball, 1990). The banks built up from their small base in the mortgage market, taking advantage of the relaxation of money-supply restrictions in the 1982 Budget. The building societies responded by redoubling their drive to attract savings by offering a range of new financial products. The increased inflow of funds enabled them to relax lending policies. They began to lend 95 per cent or 100 per cent of valuation, to increase the multiple of the amount loaned to the applicant's income and to take increasing account of second incomes in a household. The overall effect was to increase total mortgage debt in real terms by 11 per cent *per year* over the period 1980–9. The number of loans outstanding from all lenders rose from 6,210,000 to 9,628,000 over the same period – an increase of 55 per cent in debt-encumbered households in little over a decade.

According to neo-liberal beliefs about the operation of demand and supply in free markets, the effect of this rapid and sustained increase in effective demand for housing should have been to stimulate output to a commensurate degree. This worked until 1988 and then stopped working. The output of new housing from private-sector housebuilders had been about 118,000 in 1979. The subsequent record is as shown:

Table 5.2 Annual private-sector housing
completions in Great Britain 1980–92

1980	128,100
1981	114,900
1982	125,400
1983	148,100
1984	159,400
1985	156,500
1986	170,400
1987	183,700
1988	199,300
1989	179,500
1990	156,400
1991	148,200
1992	138,000

Source: NHBC, *Private Housing Statistics*

The output of the industry fell drastically just at the time when the total house-purchase credit had reached unprecedented levels. This contraction has not been caused by any shortage of materials. The suppliers of bricks, for example, have reduced their price to a fraction to try to sell them. Labour too is in plentiful supply and willing to work for low wages. In fact building costs in early 1993 were back at 1984 levels.

The combination of a large credit-based increase in demand and a reduction in the output of new properties led to a marked effect on prices:

Table 5.3 Average regional house-prices indices 1982–92 – selected regions

	1982 Q1	1983 Q1	1989 Q1	1992 Q4
UK	91.9	100.0	226.3	190.6
Greater London	91.2	100.0	278.3	188.9
Outer south-east	88.3	100.0	281.6	183.2
East Anglia	88.7	100.0	291.0	193.7
South-west	89.6	100.0	263.2	199.1
Yorks and Humber	93.8	100.0	219.2	190.0
North-west	92.9	100.0	194.2	249.1
Scotland	92.5	100.0	152.2	183.4
Northern	91.9	100.0	176.6	226.5

Source: Nationwide, *House Prices in 1992*

Plate 5.5 New homes built by McLean Homes South East Ltd in a suburb of
Eastbourne in the early 1990s. This estate of thirty-nine houses was built on a
four-acre 'infill' site bought from the Health Authority following the closure of a
small local hospital as part of the rationalisation of health services. The houses are
described as 'Victorian' and they have a strong 1900 feel with the square bays,
tile-hung finish and 'sash windows', which in fact are higher-technology than this
and tilt to open. The prices for the three- and four-bedroom houses, all of which have
an *en suite* bath, or shower room, ranged from £80,000 to £130,000. All have been
sold despite the difficult trading conditions of the early 1990s. There is a drive for
two cars; featured here is a Rover, maybe 'corporate', and a small hatchback.
Photograph by Peter Ambrose.

Prices rose sharply everywhere between 1982 and 1989. The effect was
especially marked in London, the outer south-east and East Anglia. In special
situations such as London's Docklands the rate of price inflation provided a
bonanza for those capable of buying, say, ten houses purely to sell on at a
higher price a few months later. Then quite suddenly, and following the
ending in August 1988 of double mortgage-interest tax relief if two taxpayers
were taking out a mortgage, the slide in prices began in the most 'overheated'
regions. The sharpest decline in prices occurred in London, the outer south-
east, the south-west and East Anglia. Elsewhere in Britain prices remained
steadier, and in the north-west, the northern region and Scotland prices
continued to rise even after 1989/90. It is the sharpness of the falls in the
south-east that was crucial in setting the national index, by virtue of the
weight of transactions, and that dominated media and political discussion.

This price volatility affected both new and secondhand houses. When the
prices achievable for new houses rise sharply the effect normally feeds
through via the 'residual' calculation to the sums bid for the sites on which

they are to be built. The builder works out a site evaluation by estimating the total sum he will receive on the sale of the houses, subtracting the estimated construction and related costs of the scheme and subtracting a further sum for the intended profit. He is then left with the amount he is prepared to bid for the site. If there is an anticipation that house prices are on a steeply rising curve, some at least of the effect will be felt on land prices (Ambrose, 1976; Pearce, Curry and Goodchild, 1978). High volatility in the price of this input factor will increase the risk that it may become a commodity which is tradeable in a speculative manner before being built on – an entirely unhelpful effect. Constriction of the supply of building land by the effects of the planning system is also likely to produce a price effect, but this is a matter that cannot be pursued here (see, for example, Balchin, Kieve and Bull, 1988; and Cullingworth, 1988). The incidence and effects of land-price volatility are further discussed in the English/French/Swedish comparative work reported in Chapter 10.

THE END OF THE DREAM

These sharp fluctuations in credit, housing output and house and land prices have produced, in combination, a range of problems for housebuilders, building-materials producers, construction workers and house purchasers alike. The housebuilding industry, and the materials-supply industries that depend on it, were by the late 1980s suffering the worst recession since the Second World War (Counsell, 1990) and there has been a flood of bankruptcies. This has affected old-established major companies, middle-sized firms and thousands of very small builders alike. Various 'volume' housebuilders such as Barratts, Costain, Bovis and Charles Church have declared massive pre-tax losses. Firms, such as McCarthy and Stone, which did well through the 1980s with their sheltered housing developments for the elderly find that their market has shrunk. In this case the fall in house prices means that older people trading down are realising smaller capital sums with which to purchase a sheltered home. Various companies are for sale but there are few purchasers. A more common pattern has been for the main creditors, usually the banks, to 'reconstruct' the company or for the receivers to be called in.

The medium-term prospects for the industry are not favourable either, although one survey of prospects (Credit Lyonnais Laing, 1993) believes that a limited recovery may be in sight – so long as government fiscal policy is reasonably supportive. In terms of longer-term prospects the home-ownership rate, following its long period of unbroken increase, may be near some kind of saturation level at around 70 per cent – especially since the ideology which has underpinned it has been revealed as suspect in various ways. Demographic forecasts issued in 1989 by the Department of the Environment indicate that the annual rate of household formation, which

Plate 5.6 A form of speculative construction and/or conversion for owner occupancy not discussed in the text. Following the sharp decline of activities in London's Docklands from 1967 to 1981, many of the riverside warehouses, solidly built from the eighteenth century onwards, were converted into luxury riverside flats. Initially in the early 1980s they were sold and traded for rapidly rising prices, but over-supply plus the economic difficulties of the late 1980s have led to considerable price volatility. This form of provision, although added into 'housing completions' statistics, bore little relation to the desperate housing needs of the area. These examples are on the north shore of the Thames in Wapping. The scene is graced by a rare surviving example of a spritsail Thames sailing barge. Photograph by Keith Hunt.

was close to 200,000 in the later 1980s, might fall to about three-quarters of this figure in the later 1990s (Credit Lyonnais Laing, 1991). In particular the number of married-couple households was expected to fall by 230,000 in the five years following 1991. By contrast, the number of one-person households was expected to increase by 30 per cent and of lone-parent households by 36 per cent. One-person households may well number 7.5 million by 2011 – an expectation which is engaging the interest of strategic planners in the industry.

The recent history, and especially the length and depth of the house-building recession, presents a number of vexing problems for the analyst as well as for others. The NHBC produces a 'First Time Buyers' Ability to Buy Index'. This takes into account average incomes and prices, the average deposit payable and interest rates. The higher the index, the greater the ability to buy. Since 1971 the movement of this index has been a very reliable predictor, two to three quarters in advance, of the rate of housebuilding starts. In mid-1989 the index had fallen to around 40, indicating that first-time buyers were being squeezed out of the market by the escalation of prices that had taken place in the mid-1980s (see Table 5.3). The index has since risen sharply and consistently to an unprecedented high of 110 at the beginning of 1993. On this basis things have never been better for the new entrant to owner occupancy, and buyers should be coming forward in droves. But there has been no sign of a supply response in the thirteen quarters over which the index has been rising. In other words the recovery in housing starts, based on past experience, is two and a half years late and shows no signs of coming. The housebuilding industry, often in the past a natural ally of government, is currently highly critical of the housing policies of the 1980s and early 1990s (Todd, 1992).

What has gone wrong? That question has provoked a number of high-level enquiries which have produced voluminous reports (for example Inquiry into British Housing, 1985, 1986, 1991). One problem is that the expansion of credit and the lowering of its cost by means of MITR has not been targeted solely at the purchase of new houses; it makes no distinction between the purchase of an existing house and a new one. Given that far more housing transactions relate to existing houses, the main effect of a flood of demand credit may be to increase secondhand prices, not to stimulate new output. Nor is the subsidy well targeted to those in greatest housing need. A recent study (Foster, 1993) has shown that disproportionate amounts of MITR are directed towards London and the south-east, partly because this is where prices are highest. The effect of the subsidy also appears to be regressive because in 1991/2 44 per cent of it went to households earning over £20,000, and only 15 per cent to households earning under £10,000. Two years previously the corresponding figures had been 30 per cent and 26 per cent.

But this failure to target the subsidy is a longstanding one and it does not explain the 1989–92 problems. Other factors may be advanced, the most obvious of which is the lack of buyer confidence in view of the apparent permanence of high unemployment. Although the expansion of owner occupation in the 1930s was in the context of high unemployment, the impact was felt mostly in a number of declining industries which were heavily concentrated in certain old industrial areas such as Tyneside. Most of those thrown out of work in these areas were unlikely to have been purchasing on a mortgage loan. For most of those in work, and in many other parts of the country, the 1930s was a decade of prosperity. The present effects are more pervasive in that the fear of unemployment affects those in the service industries and professions as well as those employed in manufacturing. It has also affected all regions, if to differing extents. Very few people in the early 1990s can feel totally secure about their job, and the overall level of confidence to take on long-term financial commitments is consequently low – especially since the expectation of ever-rising house prices has been undermined.

RELATED SOCIAL AND ECONOMIC PROBLEMS

The rush to take up easily available house purchase, encouraged by the belief that prices would always rise, has led to financial difficulties for millions of households. These have ranged in severity from inconveniently high monthly payments to the final disaster of repossession of the property by the lender. This means that the household is left with no home and a large debt which the sale of the house may not cover. The heavy mortgage repayments of many who bought in the high-price period, combined with the effects of rising unemployment, have led to an increase in the number of those more than six months in arrears with payments (the data are from a Council of Mortgage Lenders press release in January 1993; see also Ford, 1992):

1982	32,920
1986	65,100
1990	159,210
1992	352,050

The number of properties repossessed by lenders has risen as follows (same data source):

1982	6,860
1986	24,090
1990	43,890
1992	68,540

The figures need to be kept in proportion. In 1992 those in serious arrears amounted to 3.55 per cent of all borrowers and the repossession percentage to only 0.70 per cent of borrowers. But both situations are desperately serious for the households concerned. There is a consequent problem in that the repossessing institution needs to dispose of the property to recoup as much as possible of the loan advanced upon it. This 'overhang' of repossessed houses coming on to an already deflated market further exacerbates the problems of the housebuilding industry in finding a market for its output of new houses.

Public attention has also been focused on the related problem of 'negative equity' or the debt trap (Foster, 1993). Many of those who have purchased in recent years, taking out a loan that represented at the time of purchase perhaps 90 per cent of the purchase price, have seen the value of their property fall by more than 10 per cent so that it is now below the sum they still owe on the loan. The Bank of England estimated that about 1 million households were in this situation in August 1992 and subsequent further falls in property values may well have increased the number to 1.6 million, depending on the trend in house prices. The negative equity may amount to an average of as much as £6000 per property for recent purchasers. Initially the problem was heavily concentrated in the south-east (where prices and loans rose most sharply in the late 1980s), and the average negative equity could still be much higher in this region. But by the end of 1992 it had spread across the country and according to one estimate it was affecting one in seven of all mortgage holders (*Roof*, May/June 1993, 18). This problem, which is unprecedented, will have a number of consequences. One is that households in this situation are unlikely to try to sell their house until prices recover to the level of their debt. This in turn may well reduce labour mobility and thus detract from the overall efficiency of the economy – apart from the misery it causes to the people involved.

One effect of the house-price boom of the 1980s on public finances was to increase the cost of MITR from less than £1.5 billion in 1979/80 to £7.7 billion in 1990/1 (Oldman, 1990; Foster, 1993) – although its cost fell to £6.1 billion in the following year. This sizeable amount of tax revenues forgone has had the effect of increasing the public-sector deficit. So too of course has the cost of selling a substantial proportion of the public-housing stock at discounted prices under the 'Right to Buy' policy. Another partially hidden cost of the ideologically inspired drive to ownership has been the enormous increase in the amount of Income Support payments that has been applied to helping purchasing home owners who have found themselves unable to meet the interest payments on their loans (Income Support does not cover capital repayments). The number of such claimants has risen from 98,000 in 1979 to 411,000 in 1991, and the amount of public money spent in this way has increased from £31 million to £949 million over the same period (Foster, 1993). One study estimates the total cost of housing support in 1989/90, in terms of direct expenditure, tax concessions and discounts, at over £16 billion in capital terms and over £12 billion in revenue terms (Merrett, 1991). At least two-thirds of this sum has gone in one form or another to support home ownership. The vast cost of the pro-ownership policy has therefore contributed to the very sharp increase in the Public Sector Borrowing Requirement which has taken place – from a negative figure in the late 1980s to one of £40 billion or £50 billion in the early 1990s.

Even more fundamental arguments have been made about the relationship between the rapid rise of credit, prices and owner occupancy generally and the mounting economic difficulties confronting Britain A number of commentators (especially Muellbauer, 1990 and 1993) have argued that the rising tide of house prices in 1986–8, fuelled by the credit explosion resulting from the deregulation of financial markets, had a number of destabilising effects on the economy. In particular it allowed or encouraged people to borrow on the rising value of their properties at a time when financial institutions were competing to lend. This, as we have seen, has given rise to difficulties for many individuals and households when personal finances have worsened or interest rates risen. But not all the extra borrowing was spent on housing, and the effect on the economy was to release considerable amounts of spending on other consumer items, thus over-heating the economy. In the absence of a commensurate supply-side response this was inflationary and damaging to the balance of payments, since a great deal of the spending was on imported consumer goods. When house prices fell sharply after 1988, this pattern went into reverse, spending contracted sharply, the demand for goods and services was reduced, and many producers who had invested in increasing the supply of goods and services went out of business.

Valenca (1992) advances the additional argument that housing policy in the 1980s was partly directed towards the control of the very inflation which it seems, in the end, to have produced. If more of one's disposable income is committed to servicing a housing loan, there will be less to spend on other consumer goods. Additionally, the more people are debt-encumbered the more they are dependent on what the government does, or promises to do, with the interest rate. The promise of lower mortgage rates in the run-up to an election is obviously more effective electorally if 30–40 per cent are debt-encumbered than if only 10 per cent are. Whether this is a premeditated form of political leverage or whether it is simply a fall-out of housing policies which have some other prime aim is a matter for conjecture.

The considerable publicity given to the impact of these difficulties has produced a marked change in public attitudes to owner occupancy. Aspects of the ideology referred to earlier in this chapter have been seriously undermined. To judge by media accounts, more and more potential household formers are considering the rental option as a lower-risk way of finding at least a first home. It could well be that this change in public attitudes, which reverses the trends of the past seventy years, might be beneficial in the long term by encouraging a more balanced approach to housing issues by government and a more rounded supply response by housing providers.

THE PRESENT STRUCTURE OF THE HOUSEBUILDING INDUSTRY AND ITS PROSPECTS

There is a considerable literature on the industry (see, for example, Ball, 1983 and 1988; Merrett with Gray, 1982, chapter 11; Balchin, Kieve and Bull, 1988, chapter 10). Ball distinguished five types of housebuilder in the early 1980s – the petty capitalist who can move in and out of housebuilding; small family-capital firms who are often locality-based and need to keep their capital turning over in housebuilding; non-speculative housebuilders for whom this activity is one of a range in which they are involved; large capital firms with regional specialisation who are normally publicly quoted companies and raise equity capital but who may also be 'highly geared' and heavily dependent on borrowing; and finally the major 'volume' housebuilders. Ball discusses the characteristics, aims, strategies and mode of operation of each of these groups. Firms in the various groups are differentially at risk in recessions in the industry such as that of 1989 onwards. Survival into the next upturn depends not only on size but on capacity to diversify activities, degree of 'gearing' and fixed plant costs, quality of product and capacity to identify 'niche' markets in which demand may remain buoyant.

Survival probably also depends on the quality of marketing strategies, and in this respect some of the major housebuilders have been very innovative in the lean times. Barratts, for example, have an agency service which offers to sell an existing house at an agreed asking price for purchasers of a new Barratts house or they will take an existing house in part exchange. They also offer mortgage redundancy protection and payment of some purchase costs for first-time buyers. Fairclough Homes offer their 'Activator' service for some of their properties. Under this scheme the prospective purchaser is in fact an assured tenant for the first four years at a fixed rent, after which the purchase is made at the price ruling at the beginning of the tenancy. A mortgage is made available by one of the major banks at the point of purchase. This reduces the monthly cost over the first four years of occupancy to approximately 60 per cent of what it would have been under a conventional mortgage purchase.

By the standards of comparably large industries, the housebuilding industry is highly fragmented. According to one market survey (Credit Lyonnais Laing, 1991), the top ten firms produced an estimated output of about 41,000 units in 1991 in a total output of 148,200 – a market share of approximately 27 per cent. Twenty-two firms each produced over 1000 units, with the largest, McLean Tarmac, producing 10,200. Of the seventy largest firms, only about half have housebuilding as their major activity. Many of the rest are subsidiaries of groups which include building contractors or other activities. The survey notes that there is an increasing tendency for firms to build for 'contract' rather than speculatively. Much of the output is now to an extent commissioned – either for partnership and urban-renewal schemes, for housing associations or for BES schemes. This represents a movement away from the situation in the 1980s, when perhaps 97 per cent of houses produced by the industry were built speculatively.

This trend clearly reflects a risk-avoidance strategy. Given that the 'time to build' normally ranges between eight and twelve months, and that funds have to be committed long before the start date, much can change between the commitment of M and the realisation of M'. As we have seen, changes in financial conditions, legislation, confidence and a mass of other factors can affect the market relatively quickly. Costs too have to be accurately estimated. But some, for example the cost of borrowed money, depend largely on external factors. As one example of what can go wrong, there has normally been a charge made by the major 'utilities' for providing facilities for water, sewerage and electricity connections. This used to be reasonably related to the actual cost incurred in the work – perhaps of the order of a few hundred pounds per unit for a small estate. Suddenly in 1989, following privatisation of some of the utilities, a levy of up to £1500 per housing unit was imposed for these connections.

The experience of the 1980s has therefore both revealed and exacerbated the problems of the British housebuilding industry. It began the decade in a technologically backward and under-capitalised condition, at least when compared to similar industries in other north European countries. It first benefited then suffered from fluctuations in the supply and cost of credit, since it geared itself up to a level of demand which fell rapidly towards the end of the decade. The sharp increase in the level of unemployment, and its pervasive effects, further depressed the market by undermining purchasers' confidence about entering into long-term loan commitments. Land prices fell, reflecting falling house prices, and companies heavily involved in land speculation suffered another form of loss. In all, the trading conditions produced by a neo-liberal government proved very difficult for an industry regarded by many as its natural ally. Only as this book goes to press, in mid-1993, are there convincing signs of an upturn in fortunes. As Chapter 10 will show, the construction industry was having a much better time in interventionist Sweden.

CASE STUDY – A SMALL SPECULATIVELY BUILT ESTATE

This case study describes an actual small housebuilding company and the ways in which it was operating in the boom period of the late 1980s. The company is a private one controlled by one entrepreneur who had previously built up some capital in another business. He had carried out a number of small housing schemes over a ten-year period and identified a further development opportunity when he heard that the owner of a 1.2 acre site near one of his existing estates might be interested in selling. The site had previously been in residential use at a low density, and the vendor had obtained outline planning consent for ten semi-detached houses to be built.

The entrepreneur had considerable knowledge of the locality, and to supplement this he discussed the local housing market with two estate agents. From these he updated his knowledge of the state of the housing market in the area and especially took account of all the new properties built in recent years, their specifications and sales prices and the rate at which they had sold. At this stage he reached an 'in principle' agreement with a local bank that finance would be available, subject to a detailed examination of the new scheme's finances, up to 70 per cent of the value of the site and work in progress. The rest of the working finance was available from the sale of houses built previously.

The key step now was to calculate an upper limit on the price he could offer for the land using the 'residual' method. From his assessment of the local housing market and employment trends he estimated the selling price of the houses one year to fifteen months ahead to be £90,000 each. From previous experience, and from discussion with a quantity surveyor, he estimated the construction cost per house at £35,000. In addition he needed to make allowance for design fees, interest charges and overheads and to decide on a rate of profit required. The calculation for the land bid was as follows:

Sales value of the ten houses			£900,000
Less selling costs @ 3%			27,000
Net sales income			873,000
Less:	Construction costs for ten houses	350,000	
	Design etc. fees @ 10% of construction costs	35,000	
	Overhead and office costs	50,000	
	Bank finance costs	50,000	
	Profit @ 20% sales income	180,000	
			665,000
Available to bid for site (1.2 acres)			£208,000

In the case of larger developments, or when bidding at auction rather than carrying out a private negotiation, it may be necessary to refine this calculation in a number of ways.

Once the land sale was agreed at the bid figure the full details of the scheme, including some independent professional assessments of the costs and expected sales price and a month-by-month cash-flow projection, were discussed with the bank. Following approval, the loan funds were then made available. The developer now had to decide whether to place and manage the various sub-contracts for the materials and building work himself, to appoint a manager for this task, or to place the contract as a whole with a suitable local builder. The second course was adopted and a site manager was appointed. He placed the orders for materials and plant hire and arranged the necessary sub-contracts for the foundations, brickwork and all other necessary work on the houses.

The houses were completed according to schedule, the first being furnished as a show house. Some were sold directly on the site and some via one of the estate agents contacted earlier in the development process. All were sold at the projected price within six months of completion. In fact, subsequent to the land sale, the local planning authority allowed twelve houses to be built on the site. All other projected costs remained the same. How much profit did the builder finally take from the scheme? Had the mortgage interest rate gone up sharply so that the selling price achieved was only £80,000 (and the finance costs £60,000), how much profit would he have taken? And if he had projected these price and cost changes correctly *before* making the land bid, what would the site have been worth to him? How do rises and falls in house-price levels affect land values?

6

NON-PROFIT-SEEKING
DEVELOPMENT
– STATUTORY

This chapter will deal with development processes that are initiated by local authorities or other democratically accountable agencies (shown as box DA 1 in Figure 3.1). The motivation for the promoting organisations is to achieve some form of social utility from the development rather than to accumulate capital by showing a profit. As the development proceeds, profit from construction contracts and in other ways is derivable by other participating organisations from at least stages 2, 3 and 5. But this is not, or should not be, the main rationale for the scheme. The development process is almost invariably initiated to enable the authority to fulfil responsibilities which are statutorily defined or in some other way formally laid down by elected bodies.

The emphasis will be on 'public-sector' housing since this is the most common form of development in this category and it has until recently consumed more capital resources than any other. Moreover this is a form of housing tenure which is bound up in complex ways with other aspects of the urban process and with changes in economic strategies, social structure and political philosophies. For these reasons the history of British municipal housing has generated a rich literature which will be freely drawn upon. It is true that the amount of new housing promoted by municipalities has dropped almost to zero as a result of the policies of the 1980s. But for at least sixty years this has been a significant form of development, and with nearly 5 million housing units still in public ownership it remains as an important feature of the built environment – to say nothing of its continuing effect on social and political attitudes. Publicly promoted development activities are not confined to housing, and to illustrate this the case study included in this chapter deals with a mixed development scheme.

THE IMPETUS FOR PUBLIC-SECTOR HOUSING

The Great War of 1914–18 was a cataclysmic event which set the course of twentieth-century history. Arguably it was one of the two great pivotal points of the century (1989 being the other). British housing policy, and much else besides, was never the same again. The war drew millions of volunteers and then conscripts to the various fronts, where they were slaughtered in hundreds of thousands in disasters such as the Battle of the Somme in July–November 1916. Many came from poorer urban areas. Sometimes whole battalions, known as the 'Pals' battalions, came from the same town, and heavy losses in these units could leave a whole town sorrowing (Macdonald, 1983, 34–5). Those that survived these horrors would clearly be owed, and would demand, something better on their return to civilian life. 'Market forces' and private investment had failed to provide sufficient acceptable low-rent accommodation in the decades preceding the war (see, for example, Burnett, 1991, chapter 6). Some local authorities had invested in housing programmes, for example the London County Council had housed over 55,000 people by 1913 (Chapman, 1971). But none of this was on a sufficient scale to solve the problem and it was obvious that publicly funded interventionary policies would need to be devised.

During the war there had been considerable bitterness on Clydeside and elsewhere at the high rents charged to the thousands who flooded into the towns to work in munitions factories (Damer, 1992). The Increase of Rent and Mortgage Interest (War Restrictions) Act, 1915 was passed and set a precedent by imposing rent control on private-sector rented housing. If housing the urban masses had failed to produce an acceptable return on investment before the war, there was now little reason to suppose it would do so after. Another spur to governmental intervention was the report in 1917 of the Commission of Enquiry into Working Class Unrest (Cd 8663). Poor housing conditions were found to be an important factor behind the unrest in seven of the eight areas chosen for study. It was clear to politicians of both main parties that for this reason alone some action had to be taken. It was also clear that making good the four years' loss of housing production, plus effecting a more general improvement in housing conditions, would require state action and public money on an unprecedented scale.

THE EVOLUTION OF POLICY FROM 1916 TO 1939

The intention here is briefly to summarise some of the more important legislative and financial developments over this long period, those which have established or changed the principles and strategies of housing policy – if strategies is not too dignified a term for what has often looked more like reactive crisis management. Many fuller accounts of various parts of the history have been produced (see, for example, Addison, 1922; Bowley, 1945;

Burnett, 1991; Merrett, 1979; Orbach, 1977; Simon, 1933; and Swenarton, 1981) and there is at least one excellent 'thumbnail sketch' (Malpass, 1991). This chapter can do no more than summarise this literature. But it is important to review the main changes in policy because the standard, design and configuration of the millions of housing units that currently make up the municipally owned stock in Britain depend to a large extent on the period in which they were built and the particular subsidy regime that financed them.

From 1916 it was accepted in principle that Exchequer subsidies would be required for the task of improving post-war housing standards, and by July 1917 the Local Government Board had promised funding to local authorities (Wilding, 1973). The Board's Circular 86 of the same month invited local authorities around the country to state their estimates of post-war needs. This offer produced firm replies from only a minority of authorities. This is not surprising since it was made somewhat in a vacuum. It was not clear at this stage how the cost of the new public housing drive would be shared between central and local government. Nor was the overall strategy yet worked out. Would the aim of state intervention be simply to make good the shortage from four years of non-production and provide more housing for the 'working classes' and then let provision return to 'the market'? Or would policy go beyond that and seek a more fundamental improvement in housing standards by providing more rented housing for a wider spread of income groups?

In the last year of the war the discussion became more urgent and there were disagreements in Cabinet about how much of the cost should fall on central government. A fiery speech by Lloyd George in Manchester made a strong case for Treasury support and in October 1918 a panel of experts led by Lord Salisbury and including Beatrice Webb reported that at least 300,000 new houses would be needed after the war and that the cost of these should fall on central government. The Tudor Walters Report of 1918 (Cd 9191) went further and argued for 500,000 new units. In terms of space standards, frontages, the provision of heating and the proposed design and social mix of estates it set standards which one subsequent commentator (Powell, 1974) has shown were not always being met over fifty years later.

The General Election held in December 1918 was fought partly on the issue of 'Homes Fit for Heroes' – to use a phrase coined by Lloyd George in a speech in Wolverhampton within two weeks of the Armistice. The President of the Local Government Board, Auckland Geddes, informed the new Cabinet that the financial burden imposed on local authorities by the impending housing drive should not exceed the product of a penny rate and that the rest of the cost should be underwritten by the Treasury. This general approach was written into the Bill which Dr Christopher Addison, the subsequent president of the Board, brought forward in January 1919 and which became law as the Housing and Town Planning, etc., Act in the same year. This specified that it would be the duty of local authorities to consider local housing needs and draw up programmes to make good the shortages. Subsidies were geared so as to reduce the cost to each authority to the equivalent of a one-penny rate.These were strikingly radical financial proposals. There can have been few other pieces of legislation before or since which have allowed local authorities to embark on theoretically limitless expenditure programmes on a 'limited liability' basis and with most of the bill being covered by the Treasury.

This radical measure came about from a mix of powerful motives. To politicians like Addison it was a matter of trust, even in a sense a memorial to those who had fallen, that the wartime promises about housing improvement should be acted upon. In this way those who had suffered such dangers and hardships for their country would be rewarded with a better tomorrow (see, for example, Addison, 1922, 13). For others it was also a matter of political realities. Lloyd George warned the Cabinet on 3 March 1919:

> In a short time we might have three-quarters of Europe converted to Bolshevism.... Great Britain would hold out, but only if the people were given a sense of confidence.... We had promised them reforms time and time again but little had been done.... Even if it cost a hundred million pounds, what was that compared to the stability of the State?

A month later the parliamentary secretary to the Local Government Board argued that 'the money we are going to spend on housing is an insurance against Bolshevism and Revolution' (both quotations from Swenarton, 1981, 78–9).

What happened subsequently was well summed up in the title of the book (*The Betrayal of the Slums*) which a disappointed Addison wrote in 1922 after leaving office as the country's first Minister of Health. The task of getting a building drive of this size under way was fraught with problems. Builders' tenders for the new local authority contracts had to be vetted by the Ministry of Health (which the Local Government Board had become in June 1919). The scale of the reductions negotiated by the Ministry tended to make work for councils less profitable than privately promoted work. Land acquisition appeared not to be a problem (Merrett, 1979, 37) but there was difficulty in attracting labour to public-sector work, and raising sufficient loan finance also held up progress. Despite this local authorities built about 170,000 houses under the provisions of the Addison scheme by the end of 1923 – an impressive number but well short of the promises of 1918/19.

The fundamental problem lay in the cost of the programme. The cost to the Exchequer of the 'open-ended' subsidy arrangements rose alarmingly, as did the amount of capital borrowing that was needed for the building programmes. It was felt by some that the generous supply-side housing subsidies had encouraged profligacy in some local authorities and that building-material suppliers had increased prices in an unwarranted manner. In the economy at large an initial post-war boom was followed in 1920/1 by a serious slump. The unions suffered a number of decisive defeats, including the abandonment of plans for a General Strike in April 1921. Unemployment rose to nearly 2 million in June 1921 and unemployment benefit began to consume increasing amounts of public revenue. This forced a re-evaluation of priorities and in July 1921, as part of a round of spending cuts, the Addison arrangements were set aside by a Finance Committee presided over by the Prime Minister. Addison received the news in a letter marked 'For information'. He protested, but to no avail, and soon afterwards resigned (Addison, 1922, 29–30).

This two-year episode illustrates some recurring themes in British housing history. The brave pronouncements of 1918, made in a flood of relief and gratitude for national delivery, led directly to an Act that opened the way to a long-term solution to the housing problem. But the cost would clearly be enormous. Not that the actual annual cost to the Treasury in terms of subsidy ever approached the £100 million that Lloyd George was apparently prepared to spend to buy 'the stability of the state'. The problem was that shorter-term crises, for example rising unemployment, imposed more immediate costs that had to be met. Threats to the stability of the state come with different degrees of urgency, and expenditure on the Addison scheme had to make way for the management of more acute crises.

Some of the very strong arguments advanced in favour of expenditure on housing during the course of these early post-war years have since been lost sight of, at least in the political arena. For that reason they are well worth consideration today. The author of the 1919 Act argued that a well-funded building drive would drastically reduce the number of unemployed construction workers and thus save on benefit payments (Addison, 1922). From personal observation of cases in Wakefield he saw the extent of the housing problem – for example an ex-soldier living with his family in one room at a rent of 7s (35p) per week. Addison stressed the association between poor housing conditions and tuberculosis and pointed out that the sanatorium beds occupied by sufferers, and the cost of maintenance of dependants, stored up extra expenses for the state. With data from Glasgow, Birmingham and elsewhere he reported on the association between poor housing conditions on the one hand and high general and infant mortality, sleep deprivation, and poor health in children on the other. He even attempted (ibid., chapter 7) to assess the costs of the 'preventable' illness brought about partly by poor housing conditions and to point out the clear connection between the shortage of decent housing and industrial unrest, with all its attendant disruption and economic costs. Finally (ibid., 114) he showed that while annual expenditure on 'defence' was £4 18s 5d (£4.92) per head of population that on housing was only 5s 2d (26p). Addison's arguments have been summarised in some detail because they are still powerfully relevant in the 1990s – although for tuberculosis read hypothermia, bronchial conditions and a range of other ailments. The curtailment of the Addison scheme dissolved the housing-provision partnership between state and local authorities. The result, in the case of London, was that the intended five-year programme for 29,000 new dwellings was reduced to 8,820 (London County Council, 1928) and a significant attempt to solve the housing problem in London was nipped in the bud.

There followed after two years the Housing, etc., Act, 1923, brought forward by Neville Chamberlain, the Minister of Health in the newly elected Conservative government. Now, four years after the end of the war, the political and economic climate was quite different and the approach to subsidy more orthodox. The Act provided for a flat-rate annual subsidy from the Treasury of £6 per dwelling for a twenty-year timescale for housing units started before June 1926. The dwellings could then be either rented or sold. The Act also contained provisions for local authorities to stimulate private-sector construction in various ways, such as by grants or loans to both builders and purchasers. The expectation behind the legislation was that within a few years housing production would again become a 'free-market' activity – in fact section 6 of the Act provided that schemes would be approved by the Minister only if he or she were convinced that private enterprise would not solve the problem locally. During the lifetime of the Act it became apparent that this expectation would not be fulfilled and the subsidy was subsequently extended to 1929, although at the lower rate of £4 per dwelling. The Chamberlain Act produced a total of 438,000 houses over its six-year life – only 75,000 of them promoted by local authorities (see Burnett, 1991, 231 and Bowley, 1945, chapter 3). The other 363,000 marked, in a sense, the beginnings of the mass expansion of owner occupancy facilitated by rising real incomes for those in work, the expansion of building society lending, and the falling prices of new housing – a product partly of the supply-side subsidy and the drift down from Tudor Walters standards of design and construction.

In 1924 Britain's first Labour government came to power – albeit with a minority in the House of Commons. The new Minister of Health, John Wheatley, came from 'Red Clydeside', which had been one of the scenes of bitter protest about housing conditions during the war years (see Milton, 1973). Significantly the current financial and political realities precluded any return to the Addison principles. The Housing (Financial Provisions) Act, 1924 provided for a flat-rate subsidy to local authorities but at the higher rate of £9 per dwelling – or £12 10s (£12.50) in rural areas – payable for forty years. This was reduced to £7 10s (£7.50) in 1927 and abolished altogether by the 1933 Housing (Financial Provisions) Act. Wheatley had intended that the arrangements would remain in force until October 1939 – an uncannily accurate prediction of the termination of the inter-war period.

The housing produced was to be for renting and it was intended that rents should be fixed in relation to the controlled rents of pre-war housing, thus making them, in intention at least, genuinely within reach of the 'working classes'. But rent policy was, perhaps inevitably, left vague. The annual contribution of local authorities per unit was not to exceed £4 10s (£4.50) (Bowley, 1945, 43). Thus it can be seen that central government was covering two-thirds or more of the net annual cost of housing schemes. The principle of 'limited liability' for local authorities was preserved since if losses threatened they could be covered by increases in rents. Future losses could easily arise if construction and/or maintenance costs rose sharply. Perhaps more significantly, it should be remembered that the capital costs of the housing programmes were still to be covered by borrowing – mostly from the private finance market. The state could not possibly underwrite in advance the cost increases that would arise from rises in interest rates, and these could be recouped only from the rents.

Over 508,000 dwellings were built during the nine-year life of the Wheatley arrangements, all but 15,000 of them promoted by local author-ities. This represents something over one-fifth of all the housing built in England and Wales between 1919 and 1934. It may be seen as a significant contribution to the provision of rented housing – but perhaps not in the context of the heady 1918 Tudor Walters call for 500,000 units over a much shorter timespan. Nor were the rents that resulted from these arrangements within the means of anyone much below medium income. In fact in a detailed study of the effects of the various pieces of legislation on the provision of housing in Bristol (Bassett, 1976) it was shown that for most of the inter-war period the rents charged for the growing stock of council housing precluded all but a small proportion of what might be called 'working-class' house-holds.

The Housing Act, 1925 was a consolidating act which focused attention mainly on the inadequacies of existing housing but introduced no new subsidy regimes of any importance. Its five sections defined and confirmed the responsibilities of local authorities, defined 'unhealthy areas' and 'insanitary houses' and provided for action towards them, allowed new powers for borrowing and the issue of housing bonds, and prohibited 'back to back' housing and sleeping in cellar rooms unless they complied with certain provisions. It is safe to say that the Act did not materially improve general housing standards, but it did in a sense prefigure the next significant housing legislation, produced by Walter Greenwood, Minister of Health in the second Labour government. This Act, the Housing Act, 1930, dis-appointed some housing advocates (see, for example, Simon, 1933, 35) by its almost exclusive concern with slum clearance. Ministry of Health Annual Reports were making it clear that the housing situation in poorer industrial areas had if anything worsened since 1918 (National Housing and Town Planning Council, 1929), when there had been relatively speaking a surplus of cheap rentable accommodation. Certainly housing output had been high in the intervening period, but much of it had been speculatively built housing for sale to middle-income people in regular employment. The output of local authority rented housing had fallen as the Wheatley subsidy had been cut and economic recession had loomed in the late 1920s.

The new Act offered a subsidy to local authorities related to the number of people displaced in slum-clearance programmes. The subsidy was £2 5s (£2.25) annually per person (£3 10s (£3.50) – if the costs were unusually high) for a forty-year timespan, and the local authority's contribution was £3 15s (£3.75) per new house or flat built. There appears to have been some doubt about whether the subsidy depended on the same people being re-housed as had been displaced (Simon, 1933, 39). The structure of rents and any rebate schemes were to be at the authority's discretion but rents were generally expected to be within tenants' means. By directing the measure firmly towards slum clearance the Labour government implicitly made it clear that they expected the housing needs of all except slum dwellers to be met by free enterprise construction. Because 'slum' is an inexact concept, cities varied widely in the use they made of the scheme. Newcastle cleared only 1000 of its 61,000 houses, and the London County Council a not much higher percentage. Leeds by contrast used the measure to demolish and replace nearly one-quarter of its housing stock (Burnett, 1991, 244).

International crises played a part in the remainder of pre-1939 housing history. In the summer of 1931 the recession that began two years earlier in the United States was deepening all over Europe. Sterling was under intense pressure and the Governor of the Bank of England warned that it would not be supported by international financiers unless public expenditure was drastically cut. The Labour Cabinet was split on the issue and in August a 'National Government' was formed under the former Labour prime minister Ramsay MacDonald, with Neville Chamberlain once more Minister of Health. The general drift of policy under Chamberlain and his successor, Hilton Young, was to reduce housing subsidies to the minimum, to allow and encourage lower space standards in new housing, to confine municipal action to slum clearance and to look to private enterprise for all other forms of housing provision. The Housing (Financial Provisions) Act, 1933 abolished the Wheatley subsidy and introduced a weak scheme to encourage building societies to lend to investors willing to build properties to be made available for low rents. Not surprisingly this had little effect.

The main innovation of the mid-1930s was the Housing Act, 1935, which was partly in the nature of a book-keeping rationalisation. It required that each local housing authority must combine all its annual revenue operations – rents, debt charges, maintenance, subsidies, and so on – into one Housing Revenue Account, regardless of which particular subsidy scheme their various estates were built under. This gave local authorities much greater flexibility to set their rent structures. They were now able to calculate the total amount annually which must be charged as rent in order to balance their housing account. They could then set rents so as to reflect the 'use value' – perhaps in terms of location, size and desirability – of the various estates and housing units in their ownership. This 'rent pooling' arrangement meant that the rent charged in a particular estate was no longer directly related to the current revenue impact of the historic costs of building that estate, but was rather collectivised across the whole stock of the authority. This consolidation into one account also facilitated the granting of rent rebates to the poorest tenants, regardless of the particular estate they lived on. In fact rebate schemes were extremely unpopular since they involved means testing, a very emotive concept at the time, and had the effect of increasing the rents for the majority not receiving a rebate.

FROM THE SECOND WORLD WAR TO 1970

Housing history following the end of the Second World War in 1945 was in some ways similar to that following 1918. During the war morale-boosting promises were made about a better tomorrow – including the promise of better housing and environmental conditions. Housebuilding had come to a halt for the better part of six years; nearly half a million housing units had been lost in the war; the birth-rate was expected to increase and household size to continue its long-term decrease as more young people aspired to live independently of their parents. All this was foreseen and in March 1944 Churchill announced that there would be an emergency programme to build 500,000 EFM (Emergency Factory Made) prefabricated houses immediately the war ended. About 170,000 were eventually built (Pawley, 1992). They were regarded as simply a stopgap measure (although some are still being lived in) and in no way as a substitute for a full-scale housing drive. Political urgency was injected into the situation by the wave of squatting in the immediate post-war period (Bailey, 1992).

A Labour government was elected by a landslide in 1945, perhaps primarily because of the programme of welfare improvements promised in the manifesto, and Aneurin Bevan became the Minister of Health with responsibility also for housing. In 1946 he brought forward the Housing (Financial and Miscellaneous Provisions) Act. This used the Wheatley flat-rate subsidy principle but at the level of £16 10s (£16.50) per dwelling unit, payable for a sixty-year period. Local authorities would have to contribute a further £5 10s (£5.50) per unit from the local rates. The municipal housing was to be for 'general needs' – in fact the subsequent Housing Act, 1949, which also introduced improvement grants, was the first to make clear in its wording that local authority housing was no longer exclusively for 'the working classes'. This was a reversal of the pre-war philosophy that public housing was to be built only if it was clear that private enterprise would not produce a sufficient supply for lower-income people. The target of the Labour government was to build 240,000 new dwellings per year. Standards were to be higher than those of pre-war housing. The Dudley Report, produced in 1944, advocated *inter alia* more variety in dwelling type, a minimum floor area of 900 square feet for three-bedroomed houses, better equipped kitchens and bathrooms, and clearer separation between cooking and living spaces.

The main problem, as always with housing, was to provide for the promoting authorities a source of cheap, long-term borrowing. Bevan returned to a nineteenth-century solution. The Public Works Loan Board (PWLB) was first set up in 1817 (Merrett, 1979, 154–6) and provided finance on advantageous terms for a number of early permissive housing measures, notably the Housing of the Working Classes Act, 1890. Now in 1946 the Labour government funded the PWLB to provide virtually all the capital requirements for the housebuilding programme and at rates of interest which varied between 2.5 per cent and 3.0 per cent – conditions that were very promising for a large-scale housing programme. Very quickly, however, the government's intentions were frustrated by external pressures. The economic crisis of 1947 and the terms of the assistance from the US negotiated under the Marshall Plan (see Cooke, 1957) forced a reduction in public spending on 'social' programmes and a redirection of investment to foster economic recovery. As a consequence the Annual Reports of the Ministry of Health showed that loans approved from the PWLB for housing investment fell from £222 million in 1946/7 to £168 million in the following year. As at several crisis points before and since, capital expenditure on housing programmes had been curtailed partly for the politically expedient reason that people will not miss what they have not yet had. Nevertheless a total of 900,000 houses had been built by the time Labour left office in 1951.

The incoming Conservative administration set an overall target of 300,000 new houses per year and frequently exceeded this figure. In fact the number of local authority completions in 1953 and 1954 was the largest in British housing history, although quality standards were gradually relaxed. Soon the Minister of Housing, Harold Macmillan, began to place greater emphasis on the stimulation of private-sector building. The Housing Subsidies Act, 1956 abolished compulsory contributions from local authority rate funds to Housing Revenue Accounts and cut the 'general needs' housing subsidy to £10 per unit per year. The revised subsidy pattern also provided for much higher subsidies for flats – £32 for flats in four-storey blocks, rising to £50 for those in six-storey blocks, with an additional £1 15s (£1.75) for each floor higher than this. The incentive was now in place to encourage local authorities to place contracts for high-rise blocks of flats, although the effects of this were not seen on a large scale until the 1960s. Government also took steps to induce authorities to borrow increasing proportions of their capital requirements at commercial rates from the private capital market rather than allowing them to benefit from the advantageous rates available from the PWLB. Councils were left exposed to higher costs and had to choose

Plate 6.1 Mr Harold Macmillan, when Minister of Housing and Local Government in 1954, managing to look impeccably patrician while demonstrating the kind of 'oldstyle' kitchen requiring modernisation. The Conservative administration of which he formed a part was pledged to build 300,000 houses per year and frequently exceeded this figure. In fact in the year in which this photograph was taken nearly 224,000 local authority units were completed – the second highest total in British housing history. Photograph: Topham Picture Source.

between cutting back programmes and building to poorer standards. Rents inevitably drifted higher to help balance Housing Revenue Accounts since the expenditure side of these accounts increased sharply to service the costs of the increased borrowing.

The Housing Act, 1961 added to the upward pressure on rents, since it introduced a differential subsidy system which, in general terms, tended to reward those authorities which had increased rents most and penalised those which had built up their stock earlier (see Merrett, 1979, 183, for a more

Plate 6.2 The response to the Housing Subsidies Act, 1956 was a boom in the construction of flats in high-rise blocks. This lasted well into the 1960s but had become unpopular as a solution for social and other reasons by the end of the decade. This 1964 photograph shows components being off-loaded for the first large local authority scheme to use industrialised building techniques where the walls, floors and stairways and so on are pre-cast in the factory. This scheme for 562 flats to house about 1800 people was built at Woolwich in London and the units were constructed by Taylor Woodrow–Anglian. Photograph: Topham Picture Source.

precise account). The effect on the whole was to benefit Conservative local authorities. In the same year the Parker Morris Report, *Homes for Today and Tomorrow*, recommended standards for all new houses, public and private. These related primarily to improving internal living and storage space, especially in terms of kitchens, and to ensuring better heating standards. The final Conservative legislation, the Housing Act, 1964, extended the improvement-grant programmes, introduced improvement areas and, perhaps more important, set up the Housing Corporation, a central body to oversee the development of housing associations and other 'voluntary' forms of promotion. This 'third arm' of housing was to be fostered to provide a middle way – an option between the starkly polarised alternatives of council renting and owner occupancy. With the General Election looming, and with council housing completions having fallen to only 105,500 in 1961, this was

Plate 6.3 Another part of the drive to produce more public-sector housing in the mid-1960s. These 'mobile homes' in part of Lewisham were built by the Greater London Council as part of an 'overspill' scheme to house people from inner London. They were built to higher standards than the immediately post-war 'prefabs'. They are effectively two-bedroom detached houses with bedrooms of the size recommended by the Parker Morris Report. They have front and rear gardens and are thus more popular than high-rise flats for families with children. It was probably not envisaged that they would still be required thirty years later.
Photograph by Keith Hunt.

perhaps a political move both to boost output and to stimulate a form of 'social housing' which was not so clearly associated with leftish voting and for which rents could be engineered upward with less tenant opposition.

The Labour administrations of 1964–70 began with the intention of building 500,000 houses per year, half in each of the two main sectors. But it is significant that their 1965 White Paper, *The Housing Programme 1965–70* (Cmnd 2835), saw the public part of the programme as primarily to meet the exceptional needs revealed partly by media attention on the housing problem. It saw owner occupation as the 'normal' long-term form of housing provision. In terms of numbers there was considerable success, and total output exceeded 400,000 units in 1967 and 1968, the highest annual totals ever achieved in Britain. The high rate of public-sector completions, over 181,000 in 1967, was made possible by changes to the subsidy arrangements in the Housing Subsidies Act, 1967, which tied the funding more closely to the actual finance and construction costs involved. In a partial return to the Addison principles, the subsidy now took the form of the difference between actual borrowing cost and the cost had the borrowing been at 4 per cent. To avoid the 'open-ended' commitment that had caused so much Treasury grief

114

in the Addison era, building-cost yardsticks were implemented which, if exceeded, would lead to a loss of subsidy. In this Act too the subsidy arrangements which had favoured industrialised high-rise building during the early and mid-1960s were abolished and construction in this form declined – even more quickly after the partial collapse in 1968 of Ronan Point, a tower block in east London.

In terms of output, if not entirely by conviction, the later 1960s was perhaps the last period to date in which a British government accorded a high priority to housing provision using collectivised principles and with some protection from the growing exigencies of economic decline and the instability in interest rates brought about partly by sterling's increasing problems in the world's financial markets. Labour's last measure, the Housing Act, 1969, focused more on the rehabilitation of inner-city areas and increased the level of improvement grants. Local authorities received as subsidy three-eighths of the interest charges on the money borrowed to fund improvement expenditure. The Act also introduced the idea of General Improvement Areas. This, rather than large-scale programmes of local authority building, was judged to be the most that the national economic situation could sustain. This was very much to the liking of the many urban councils that swung to the Conservatives in the 1967 and 1968 local elections. But the brave attempt to solve the housing problem by a massive drive to make decent housing available at 'social' rent levels was over.

NEW APPROACHES SINCE 1970

The first important housing legislation to come from the 1970–4 Conservative government was the Housing Finance Act, 1972, following the 1971 White Paper *Fair Deal for Housing*. This was mostly about subsidies and rents. The long-term intention was to bring about a decrease in the former and a differentiated rise in the latter. The general level of rents was to be fixed at a 'fair rent' level to be brought progressively into line with private-sector rent setting as carried out under the Rent Act, 1965. Concurrently an increasing proportion of subsidy was to be used to allow rent rebates to those on low incomes. In terms of Figure 3.1 this marks a significant shift of support from stages 2 and 3 to stage 4. The 'fairness' presumably lay in the more precise targeting of support. Meanwhile the amount of house-purchase credit lent out by building societies increased from £2 billion in 1970 to £3.6 billion in 1972, private-sector housebuilding boomed, the ratio of attractiveness between purchasing and council renting rose sharply, and house prices more or less doubled in a two-year period. In 1972 also unemployment soared, there was a run on the pound followed by devaluation, interest rates rose and a freeze was imposed on wages, prices and rents – except council rents. The macro-economic difficulties continued into 1973, when OPEC oil prices were quadrupled. In all the circumstances it is not surprising that local authority housing starts in 1973 were fewer than in any year since 1945.

After this hiatus the period of Labour government from 1974 to 1979 was more strongly marked by reflection than action. Although subsidy expenditure on council housing rose sharply during this period, much of the increased spending was consumed in higher interest charges – the PWLB rate rose from 8.9 per cent in 1972 to 13.4 per cent in 1974. Much subsidy was also spent on increased rent rebates and allowances and higher land and construction costs. It is likely that the latter two costs had been sharply bid up by the boom in private speculative housebuilding. Output rose somewhat in the mid-1970s but never to the levels of the later 1960s. The Housing Act, 1974 expanded the role of the Housing Corporation, and the Housing Rents and Subsidies Act, 1975 abolished the 'fair rents' concept and set up new provisional subsidy arrangements based on pooled historic cost principles. Then a housing-finance review body was set up which deliberated for two years before producing in 1977 a Green Paper, *Housing Policy: A Consultative Document* (Cmnd 6851) – variously described as 'insipid' and 'anodyne' by later commentators. One of them (see Merrett, 1979, 269) draws attention to the paper's claim that for most people home ownership is 'a basic and natural desire'. Quite reasonably he dismisses that view as 'psychological nonsense'. This document makes it clear that the Labour government was already facing along the road shortly to be taken by the Thatcher administrations. Since by this time about 57 per cent of voters were home owners perhaps it would have been naive to expect anything else.

The Thatcher years saw both an effective centralisation of housing policy and a multi-faceted drive towards privatisation at all stages of the process shown in Figure 3.1. The Housing Act, 1980 delivered the manifesto promise of a 'right to buy' for council tenants. Thus the main policy thrust was to increase ownership by a tenure shift of the existing stock rather than by the differential addition of new owned and rented stock. This has brought about a positive shift in the B:A ratio in Figure 1.1 and means that an increasing proportion of available housing has been allocated on 'market' rather than 'social' criteria. Discounts of up to 70 per cent were offered as an incentive to purchasers, and the number of units sold quickly rose to about 200,000 in 1982, after which it fell substantially. In total about 1.5 million units have been sold, ranging from about one-fifth of the original stock in the north region to about one-third in the south-eastern. The financial and other consequences of this policy have been fully discussed elsewhere (for example Forrest and Murie, 1990). For the period up to 1989, the receipts from these sales amounted to well over £17 billion – nearly half the proceeds of all forms of privatisation. Very little of these proceeds was used for new public-housing investment, which fell in 1991/2 prices from over £12 billion in 1974/5 to £3 billion in 1991/2 (*Roof*, May/June, 1992, Housing Update). Most of the capital receipts from sales have been applied to the reduction of existing debt or of the level of new borrowing.

The other main Conservative legislation has been the Housing Act, 1988. This Act is primarily concerned with redefining landlord–tenant relationships in the private sector, with new funding mechanisms for housing associations, with the setting-up of Housing Action Trusts to take over local authority estates in some areas, and with procedures for new landlord organisations to come forward and acquire council stock by 'Change of Landlord' or by 'Voluntary Transfers' (under powers in the Housing and Planning Act, 1986). The clear aim is to run down the stock of housing owned by local authorities as quickly as possible by substituting new private- or voluntary-sector owners. Large-Scale Voluntary Transfers (LSVTs – see *Roof*, May/June, 1991, 13; and Mullins, Niner and Riseborough, 1993) produce capital receipts which help reduce pressure on the Public Sector Borrowing Requirement, but they have implications for increased revenue expenditure on Housing Benefit to cover the higher rents which inevitably result (McIntosh and Utley, 1992). In fact the average number of recipients of Housing Benefit has been between 4 and 5 million since 1986/7 and the cost has risen from about £3.5 billion in that year to an expected £8.0 billion in 1993/4 (Burrows, Phelps and Walentowicz, 1993). Finally, the Local Government and Housing Act, 1989 has set out the new financial regime for local authority housing in England and Wales (Malpass and Warburton, 1993). It takes steps to control more closely the total amount of capital borrowing and specifies that only up to 25 per cent of the receipts from the sale of housing stock can be used to fund new projects; the rest must be applied to the redemption of debt or to interest-earning deposits.

In the early 1990s, as a result of all these measures, the proportion of British housing owned by local authorities is around 20 per cent. Much of the better council stock has been sold and the remainder is beginning to look like a 'residual' form of tenure – a suitable resting place for those on state benefits and least able to compete in an otherwise marketised system. This outcome is certainly in line with the intentions of the 'neo-liberal' administrations which have overseen housing policy since 1979 and which have sought to reduce the role of local authorities from that of the main provider of social housing to that of 'enabling' provision by other agencies (Bramley, 1993). It is a long way from the progressive visions of Addison, Wheatley, Bevan or even Macmillan. But then British society has come a long way over the past seventy years.

SOME OF THE BUILDERS AND WHAT THEY BUILT

This section, which will necessarily be much shorter than the previous one, focuses on stage 3 of the model shown as Figure 3.1. Who carried out all the building under the various policies and subsidy regimes reviewed, and to what designs and residential layouts did they build?

One set of builders has been the Direct Labour Organisations (or DLOs) –

the building departments of the promoting local authorities. The first of these, the London County Council DLO, was set up in 1892 by the ruling 'Progressives', since it was judged that the private construction industry could not provide an adequate service for the newly emerging local authorities (Direct Labour Collective, 1978). Other London authorities such as Battersea and West Ham soon followed this example. In 1919 and 1920 a total of seventy new DLOs had been formed, and they built between them nearly 6000 of the new houses produced under the Addison Act. During the 1920s they continued to build an increasing proportion of council houses, predictably in the face of fierce opposition from the National Federation of Building Trades Employers. DLOs doubled in number in the years immediately after 1945 and were actively involved in the housing drive under Bevan. From 1955 to 1967 total DLO employment rose from 70,000 to 200,000, although less than half of these were engaged on new building. This was perhaps the high point of DLO activity, since many of the departments did not have the technology to become deeply involved in the 'industrialised' building methods that marked much of the 1960s. Since then their role in new building, although not maintenance, has declined steadily to almost nothing today.

An account of London County Council housing construction in the 1920s (London County Council, 1928) offers much contemporary information on activity under the early subsidy regimes. Between 1919 and 1927 the LCC had built 23,372 houses and flats. Of these 12,130 were on the Becontree (Dagenham) estate, 2096 at Bellingham and 3,225 at Downham – so almost three-quarters of the building was on the three biggest estates. The first two of these were commenced in 1920 under the Addison arrangements and continued under the Chamberlain and Wheatley Acts. In view of the size of contract involved in such large-scale projects they were awarded to well-established builders with a capacity to do both building and infrastructural work. Given the instability in prices and wage rates in the immediate post-war years, and the possibility that changes of layout might be required as the work proceeded, the contracts were awarded on 'cost plus contractor's fee' basis rather than by tender or by fixed-price contract.

The estate at Becontree was built on a 3000-acre site acquired by compulsory purchase. This was enough for 24,000 houses with additional sites to be leased for private development. At prevailing average household sizes the expectation was that this would be equivalent to building a town of 130,000 people – the equivalent of Brighton at that time. The construction contract for this immense undertaking was placed with C. J. Wills and Sons Ltd. Provision was made for churches, schools, shops, parks and public buildings. A large-scale employer in the shape of the Ford Motor Company came to nearby Dagenham in 1931. The population reached 90,000 in 1934, which made Becontree the largest planned suburb in the world. The estate consisted almost entirely of two-storey houses or 'cottages', some terraced and some arranged in groups of four.

Plate 6.4 A leafy view of part of the Bellingham Estate in south-east London. This estate was approved in July 1919 by the London County Council using the Addison subsidy scheme. The contract was placed with McAlpines in 1920, and over 2000 units had been built by the end of 1922. The roads, many of them tree-lined, radiate out from the small central park from which this photograph was taken. Several churches, shops and a pub were built to serve the needs of the residents, and a nearby station gives easy rail access to central London. Photograph by Keith Hunt.

The Bellingham Estate in Lewisham was approved by the LCC in July 1919 as a scheme to accommodate at least 8750 people. A 252-acre site was purchased and some sites leased for shops and the building of the 'Fellowship Inn' adjacent to Bellingham station. A contract was placed with Sir Robert McAlpine and Sons in October 1920, and in little over two years 2090 houses and flats had been built – 403 five-room houses, 111 four-room 'parlour' houses, 1100 four-room non-parlour houses, 188 three-room houses, 156 three-room flats and 132 two-room flats. The estate was planned with roads radiating out from a small central park and within a few years two churches had been added. The total land and development cost for the estate was £2,358,000 – an average of £1128 per unit. The estate must have been a healthy one – it produced Henry Cooper.

The third large LCC estate, that at Downham, was built on a 522-acre site of which over fifty acres were reserved for open space. This estate was built under the Wheatley Housing Act, 1924, which, following Tudor Walters recommendations, specified a permitted maximum density of twelve houses per acre. This was a huge improvement on much existing urban housing which might be at fifty houses or more per acre. Nearly all the building was of two-storey cottages, with a few three-storey blocks of flats. By the end of

Plate 6.5 Part of the Downham Estate promoted by the London County Council under the provisions of the Wheatley Act, 1924, which provided for a forty-year 'flat-rate' subsidy for each unit. The estate was built on the south-eastern fringes of London and the population grew to over 12,000 within three years. The maximum permitted density was twelve houses per acre – there are back gardens as well as spacious front gardens. This density contrasts to the fifty or more houses per acre which would have been the pre-1914 norm for middle- to lower-income people in inner urban areas – one measurable gain for the heroism and blood expended in the Great War. Photograph by Keith Hunt.

Plate 6.6 Once sold to private owners, the Downham Estate houses can undergo dramatic transformations. Here Georgian windows, shutters, a porch, rough-cast rendering, timber framing and foliage round the door have turned an original centre terrace house into what might be termed 'LCC stockbroker' style. Photograph by Keith Hunt.

120

1927 a total of 3225 units had been completed and the output was averaging thirty-nine houses per week. Sites were set aside for shops, schools, churches and 'a licensed refreshment house'.

Many smaller estates were built by the LCC during the period up to 1927. One at Norbury (which had been begun before the war) was of 218 houses developed at a total cost of £232,000 – an average cost of £1064 each. Another was the Old Oak Estate at Hammersmith: 736 houses developed at an average cost of £954. A third example was the celebrated Roehampton Estate in Putney, developed on a sloping 147-acre site. Five different contractors, one of whom went bankrupt, were used over a six-year period. The average development cost of the 1212 dwellings, which were of varied and interesting designs, was £1154 each and part of the site was subsequently leased to a sixth contractor for the building of larger private houses – an interesting early example of mixed development. Not all the early local authority building met the high standards shown by LCC estates. Seebohm Rowntree was very critical of the unimaginative way in which York had built over 3200 houses on various estates around the city, often on a grid-iron street pattern.

As we have seen, Greenwood's Housing Act, 1930 inaugurated an era of slum clearance and urban redevelopment rather than the development of low-density 'cottage' estates on greenfield sites. Increasingly, and especially after the ending of the Wheatley subsidies in 1933, local authority housing took the form of blocks of flats, the updated version of the traditional tenement. Flats were favoured in redevelopment situations because a greater proportion of existing residents could thus be rehoused on the site and because they attracted a special subsidy under the Greenwood Act. By 1936 the LCC was building more flats than houses, and both Manchester and Liverpool had embarked on ambitious programmes which were interrupted by the war.

The classic example of pre-war flat building, in which social idealism (rooted partly in Parisian and Viennese workers' housing schemes), modernist architecture and inadequately tested technology were combined in equal measure, was the Quarry Hill flats in Leeds (fully discussed in Ravetz, 1974). By 1933 the city had made little use of the Greenwood legislation and still had some 75,000 'back-to-back' houses. A new Labour council appointed a City Architect and briefed him to build a 'state of the art' flat complex on a cleared 26-acre site not half a mile from the Town Hall. The first of the 938 flats was not ready for occupancy until 1938 and the complex not fully populated until well into the war years. This attempt to use a 'model estate' to generate 'community by design' began to go sour in the 1950s, perhaps as part of the general loss of social collectivism which occurred after the war, and many people wished to leave the estate. Considerable structural renovation and improvement were necessary in the 1960s, little more than twenty years after construction. Some of this had the effect of removing garden space and playgrounds and producing a more sterile environment. In

January 1973 the decision was taken to demolish the flats. The story had been on the whole a sorry one and provides ample evidence for the weakness and inflexibility of this rather utopian approach to the provision of decent public rented housing.

Despite the clear evidence that 'cottage' estates of two-storey houses were more popular than schemes such as Quarry Hill, the early 1960s saw a dramatic increase in high-rise building. Whereas in the late 1950s only 6.9 per cent of local authority housing contracts were for blocks of five storeys or more, by 1966 the proportion had risen to 25.7 per cent. Blocks of fifteen storeys or more represented over 10 per cent of contracts in 1965 and 1966 (Cooney, 1974). One reason for this lay in architectural ideals both about the visual punctuation of the built environment and about 'community living'. Another lay in the pressures from both central and local government in the later 1950s to redevelop older urban areas in ways which would produce most dwelling units per funds invested and land used and which would thus minimise the 'overspill' of population from redeveloped areas. This solution was still being pushed by the Ministry of Housing and Local Government in the paper *Residential Areas: Higher Densities* (HMSO, 1962) some years after the arguments had been shown to have very shaky foundations (see, for example, Stone, 1959).

But by the late 1950s – with an eye on the Housing Subsidies Act, 1956, which provided for subsidies that increased with storey height (see Merrett, 1979, Appendix 1) – a number of contracting firms with civil-engineering capability were investing in new high-rise construction techniques. Given that such industrialised methods required less on-site skilled labour, at that time seen as a bottleneck, and that they were supported by the newly set up National Building Agency, the 1960s high-rise boom is not difficult to explain. But following the growing sensitivity to the adverse criticism of high flats, especially for families with children (Maizels, 1961), a belated understanding of the real economic cost of building high and finally the wave of antipathy generated by the collapse of Ronan Point in 1968, the movement away from this solution was rapid. By 1970 only 9.8 per cent of new housing contracts were for blocks of five storeys or more. The social and financial legacy of the high-rise boom is still being felt. One approach to the problem is to replace existing high-rise blocks by low-rise blocks and houses. This is occurring in the London Borough of Waltham Forest by means of a £170 million programme operated via a Housing Action Trust and with full community involvement (Owens, 1992). The initiative is highly regarded but it should not really have been necessary to spend funds of this magnitude on refurbishing housing built only about thirty years ago.

Following the rise and fall of high-rise, the late 1960s and early 1970s saw an increasing interest in high-density but low-rise schemes. The old idea of large-scale, twelve houses to the acre, urban fringe developments became outmoded as ecological and infrastructural cost considerations alike dictated tighter patterns of development. Many of the new ideals were incorporated in the *Essex Design Guide for Residential Areas* (Essex County Council, 1973), produced for private and public developers alike. It advocated 'stepped' and 'staggered' terraced housing, arranged in 'village-like' layouts of courts and mews, and at densities of seventeen or so to the acre. Rear gardens only were encouraged. Some critics made the point that the recommendations in the guide might be more appropriate in urban infill schemes than in village or rural situations.

By the later 1970s, as we have seen, a combination of public-sector spending constraint and Labour government lack of interest had much reduced the funding and the political energy behind the development of local authority housing schemes. The advent of the neo-liberal right in 1979 has almost extinguished such energy as was left. The number of dwellings completed by local authorities in England and Wales, which even in the mid-1970s stood at 129,000, fell to 87,800 in 1978, to 30,200 four years later and to 16,300 in 1988. In the prevailing climate, riddled with central government disincentives and constraints, few local authorities can contemplate building programmes of any size. After a long and chequered history that began in a wave of national gratitude in 1919, public-sector housing production has ceased to be a significant form of addition to the built environment.

CASE STUDY – WATNEY MARKET

In this case study a local authority is promoting a mixed-use redevelopment and acting as an 'enabler' to achieve housing gain and improvement. The development is situated in the Wapping neighbourhood of the London Borough of Tower Hamlets (see Chapter 2) a few hundred yards outside the LDDC area. In the late 1960s a GLC scheme was built comprising two 25-storey steel-framed blocks of flats, two parallel rows of shops with maisonettes above, space for market stalls between (the area had previously been a street market), associated low-rise housing, and underground parking. The Borough is the freeholder of the site and manages the shop units. A number of problems have now become apparent. The high-rise blocks are unpopular, full of asbestos and costly to maintain, the mix of housing units does not suit local needs, the underground car park has become a security risk, and the main retailing 'magnet', the Sainsbury supermarket, is likely to be moving to a larger site. This would decrease the viability both of the other shops and of the market stalls.

The scheme is scheduled to be carried out over a six-year period starting in 1993 and will involve a number of private contractors. Tenants are involved in decision-making by means of regular meetings and are kept informed by a newsletter and 'surgeries' on the site. One tower block is being dismantled, which involves the careful removal of large amounts of asbestos. The other block is being remodelled externally with the addition of a

Plate 6.7 An aerial photograph, taken from the south-east, of the Watney Market development before the refurbishment work referred to in the Case Study. The parallel low-rise retail and residential units can be seen between the two tower blocks. The left-hand block is being demolished and the areas north and south of its site will be used for low-rise housing. The area south of the site will be extensively landscaped and used as a community garden. The Commercial Road can be seen in the top half of the photograph and the Docklands Light Railway crosses the bottom. Photograph: Aerofilms.

surrounding thick wall of brickwork to increase strength, reduce maintenance costs and give a more 'solid' appearance. Internally it is being rearranged to increase the number of flats from 95 to 138, most of them smaller than the existing ones.

Additional low-rise housing is to be built. Some of this will be on sites to be transferred free to housing associations, including one which will provide accommodation for people with mental and physical disabilities (see Chapter 7). The associations will receive the rents and maintain the properties but the Borough will retain the right to nominate tenants. A further area is to be sold for private housing development at a later stage. The overall effect of the

124

scheme will be to increase the number of housing units from 324 to 373 and to achieve a better mix both of unit size and of tenures. Security will be much improved and the general appearance will be softened by planting trees and shrubs.

The total scheme is costed at about £21.2 million (1993 prices). About £15.7 million will be made available under the Department of the Environment's Estate Action Programme, £1.1 million by the Borough, £2.2 million as Housing Association Grant (see Chapter 7) and £2.1 million from private sources, partly from the sale of the land for the private housing. It is important to note that the term 'made available' does not mean 'granted'. It means that either Basic or Supplementary Credit Approvals have been or will be made to borrow these sums, largely from the private capital market (see Figure 3.1). The Borough will have to pay the interest and repayments costs over whatever term is agreed with the lenders – probably 30–40 years. It should also be noted that the ratio of 'public' to 'private' finances in this scheme is higher than average, perhaps in recognition of the severe problems evident in this area of London. But, as was argued earlier (see Chapter 3), the 'public–private' distinction is misleading anyway, since 'public' borrowing depends on 'private' lending.

7

NON-PROFIT-SEEKING
DEVELOPMENT
– VOLUNTARY

This chapter deals with a wide variety of development processes which all
share two broad characteristics. They are initiated by organisations in the
NDA part of the model shown as Figure 3.1, but unlike other development
promoted in Box NDA 1 they are undertaken primarily on a 'non-profit'
basis. They require a separate chapter because they are neither 'public' in the
sense of state-provided nor 'private' in the sense of capital-accumulative.
They do, however, depend crucially on both public agencies and profit-
seeking organisations at all stages of the development process.

  Their promotional activity (stage 1) is increasingly 'enabled' and supported by local authorities, their funding (stage 2) comes from a combination
of government grants and private-capital market loans, and the construction
work, whether new-build, conversion or maintenance (stages 3 and 5), is
normally undertaken by private contractors. The form of development they
produce is mostly housing for rent, although some is for sale. Following the
policy changes of the 1980s they are regarded by government as the main
source of provision of 'social housing' – in fact the percentage of all so-called
'public-sector' housing that is produced by housing associations has risen
from 3 per cent in 1975/6 to 76 per cent in 1991/2 (*Roof*, March/April, 1992,
11). They normally exercise the right to allocate a proportion of the tenancies
for the housing they produce and manage (stage 4) while typically the local
authority will have power to allocate the rest. The development processes
they carry out are often collectively described as 'voluntary-sector' because
they are initiated not as a result of some statutory requirement but by the
voluntary action of a group, or even initially an individual, with no formal
democratic status or accountability. Incidentally it should be noted in passing
that senior people in the sector are themselves unsure how best to categorise
their activity – although very few see it as 'voluntary' (Dwelly, 1992).

  In a book organised around the themes of urban process and power the
apparent lack of democratic accountability will clearly come under some
scrutiny. Without pre-empting that discussion, which will fall in Part IV of

126

the book, it is as well to observe at this stage that in terms of user sensitivity and management participation, provision for groups who might otherwise be lost sight of, and innovation in housing design, construction and management the voluntary sector looks increasingly as if it has something very useful to offer. This chapter will include some examples of voluntary-sector work in Britain, while reference to some aspects of the work of housing associations and co-operatives in Sweden and Denmark will be made in Chapter 10.

VOLUNTARY-SECTOR HOUSING FROM THE EARLIEST DAYS TO 1919

Housing associations in Britain are thought to have their origins in the 1840s (White, 1992) – although one member association in the National Federation of Housing Associations was founded in 1235 and Church-owned accommodation for the old and needy in the form of almshouses goes back much further than this. The 1840s was a decade of hardship and political protest in many countries of Europe, of famine in Ireland and of Chartism in England. The protest of the Chartists was comparatively muted. As one social historian has pointed out (Jones, 1991), while half a dozen European capitals were in the hands of revolutionaries the Chartists hired cabs to go to Parliament to present their petition.

At this time there was fear and concern among the rich and powerful on a number of grounds. The entrepreneurs who had developed housing for rent in the rapidly growing urban centres had naturally built at high densities to economise on land costs and had filled their properties with as many rent-paying tenants as possible. In the 1830s and 1840s some of the consequences of high-density occupancy without proper infrastructural arrangements were becoming apparent. In Edinburgh the environmental effects of the 'Foul Burn', effectively an open sewer, were so pervasive that in 1832 the Commissioners of Police promoted a private bill to end the nuisance (Ashworth, 1954, 55). A Royal Commission of 1844/5 made a number of radical recommendations about future building standards and mechanisms for public inspection (Royal Commission on the State of Towns and Populous Districts, 1845). A mass of information was emerging about urban conditions, and the cholera epidemic that swept across Europe in the 1840s lent urgency to the cause. Not only were the unhealthy conditions carrying costs for the economy (Ashworth, 1954, 54); they were also widely regarded as possible breeding-grounds for radical political activity. Apart from these reasons for concern there were genuinely philanthropic motives at work in the minds of a number of reformers.

The reaction to these conditions and the dangers they represented took various forms – many of them with 'sanitary' overtones. The Metropolitan Association for Improving the Dwellings of the Industrious Classes, often seen as the first modern housing association, was formed in 1841. It was followed by the Society for Improving the Conditions of the Labouring Classes, of which Prince Albert consented to be president, thus beginning a tradition of royal interest in housing issues which is still very much alive. Both associations began to build in parts of London during the 1840s. Further associations followed in the 1860s, including the Improved Industrial Dwellings Company and the Peabody Trust – financed by a donation of £150,000 from the American George Peabody in 1862. The motivations of those involved were partly altruistic, but partly too there was a desire to show that healthier housing could be produced for the labouring classes at a development cost that yielded an acceptable return on capital invested. Ashworth quotes research (Ashworth, 1954, 83) showing that the first block of dwellings of the Metropolitan Association was producing a profit of 5.25 per cent in 1870 and that the first development of the Improved Industrial Dwellings Company produced an 8 per cent annual profit – better than most other forms of investment available at the time.

By the mid-1880s, the decade in which the East End Dwellings Company, the Four Per Cent Industrial Dwellings Company and the Guinness Trust were formed, it was estimated that over twenty-eight associations were in operation which between them had housed over 32,000 people in London. By 1905 the nine largest housed 123,000; by 1914 housing trusts alone managed over 50,000 units. In all it is clear that up to 1919 it was the voluntary associations, rather than local authorities working under permissive legislation, that were making the largest attack on poor housing conditions in London (outside London the voluntary housing movement was far less important). It should be noted, however, that although the associations had access under the Labouring Classes' Dwelling Houses Act, 1866 to advantageous loans from the Public Works Loan Commissioners (up to forty years' repayment and around 4 per cent interest), and that the trusts benefited from charitable donations, neither could provide housing at rents that the poorest could afford. It was very much housing for those in regular work, not for the unemployed and destitute who together made up 'outcast London' (Mearns, 1883). It was the urban conditions experienced by millions who could not afford the 'charitable' rents in the voluntary sector that provoked the changes in policy described in the previous chapter.

In passing it is worth noting the attempts of private individuals in the mid-nineteenth century to improve the housing conditions of the masses. Most celebrated among these was Octavia Hill (see Darley, 1990; and Clapham, 1991). In 1864, seeing at first hand the unhealthy basement in which one of her handicraft students lived (Bell, 1942), she felt moved to do something. Fortuitously her friend and tutor, John Ruskin, had been left some money by his father. Ruskin agreed to her idea to invest in 'a small lodging house, where I may know everyone and do something towards making their lives healthier and happier'. Faulty syntax apart, this seems an irreproachable Victorian statement incorporating idealism, action and maternalism in equal doses. By the spring of 1865, Ruskin's money had been invested in three houses in Nottingham Place, Marylebone. Both participants in the scheme agreed, for subtly different reasons, that the rents should be set to produce a 5 per cent return on the capital invested; she because the intended tenants would be 'working men' who 'ought to be able to pay' for their accommodation, and he because such a rate of return from such a worthy activity might have an exemplary effect on other investors.

Within a few years Octavia Hill had bought a number of other properties in varying states of dilapidation and with heavy rent arrears. She 'turned them around' as one would say nowadays, getting the common areas cleaned, reducing densities, organising repairs and taking a very firm line on the prompt payment of rents (Beatrice Webb helped collect rents for a period). In many ways, and perhaps disregarding the 'Lady Bountiful' aspects, there is much the modern housing investor and manager can learn from Octavia Hill. Her meticulous 'hands-on' approach produced materially better conditions for her tenants and a 5 per cent return on capital (£48 on £838) within eighteen months (Bell, 1942, 97). She went on to bigger, if not better, things in the ensuing decades, becoming involved in university 'settlements' in the East End, co-founding the National Trust in 1895 and working to secure much-needed parks for Londoners (on the tennis courts of one such, in New Cross, the present author developed a mean backhand as a lad).

Some employers, too, by the mid-nineteenth century were realising that investment in decent housing for their workforce might well make economic sense in the middle or long term. Ill-health, apathy and lack of sleep stemming from poor and overcrowded housing conditions did not add up to a keen and effective labour force. Perhaps the earliest example was the houses built and let at low rents by the industrialist David Dale at New Lanark in Scotland in 1784. Robert Owen came in 1799, enlarged the village, invested in the housing and community facilities, increased the factory output and generally demonstrated that a good environment could lead to increased labour productivity. Other proposals for 'new model villages' were advanced by Buckingham in 1849, the firm of Messrs Richardson at Bessbrook in Ireland in 1846, and Price's candle factory at Bromborough in England in 1853 (Ashworth, 1954, chapter 5). In the twenty years following 1851 Titus Salt built the town of Saltaire adjacent to his new factory on the banks of the River Aire near Bradford. The town had a very wide range of urban facilities from wash-houses to an infirmary – but no pubs. George Cadbury followed Salt's example at Bournville, although not exclusively for his own employees, and the Lever brothers' development of Port Sunlight is another example of 'voluntary' housing development as an adjunct to industrial activity. It is safe to assume that these attempts to produce a better living environment using 'garden village' principles were undertaken for a mixture of motives that probably included 'pure' philanthropy but certainly did not exclude the desire to secure a more dedicated and healthy workforce (see also Gauldie, 1974, chapter 16).

But all this well-intentioned voluntary activity was as a drop in the ocean when compared to the scale of the urban housing problem around the end of the Victorian era. Attitudes to the problem were far from consensual. When in 1851 Henry Mayhew produced his pioneering social survey *London Labour and the London Poor* (see Yeo and Thompson, 1971) his methods owed rather more to journalism than to social science and he portrayed the London poor as somehow not English – they had more in common with the 'inferior' races then being 'discovered' in Africa. But by 1890, when General William Booth of the Salvation Army published his plea for social reform (Booth, 1890), he felt that the wealth available in 'Greater England' would be sufficient to rescue the 'submerged tenth' of 'Darkest England' if more fairly distributed. On another tack the eugenicists, encouraged by Galton's *Hereditary Genius* published in 1869, felt that the problem was one of inadequate people and could be solved by selective breeding.

Charles Booth, a Liverpool businessman and a Conservative, felt that reports of the extent of poverty had been exaggerated and initiated his own surveys. He devoted a lot of money and much of the rest of his life to establishing the extent of poverty in late Victorian England (Spicker, 1992). He found that it had in fact been understated, not overstated – over 30 per cent were living below his defined poverty line (Simey, 1960). Finally Seebohm Rowntree, impressed by Charles Booth's work, began his own survey of poverty in York in 1899 and in other cities subsequently. Using more sophisticated methods, he too concluded that over a quarter of the population were living in poverty (Briggs, 1961). Poor housing conditions were a central element in this morass of poverty and by 1919 no effective solution had been found. The reason was as simple then as it is now – there was no profit, let alone a 5 per cent return on capital invested, in providing housing for the very poor.

FROM 1919 TO 1961

In the decades between 1919 and 1939 there were few significant developments in the voluntary sector. The Housing Act, 1919 did allow for the payment of subsidies to housing trusts, and appeals were made to philanthropic organisations for low-interest loans and gifts. Voluntary associations carried out a variety of tasks, managing older urban property, co-operating with local authorities in slum rehousing schemes and carrying out surveys to demonstrate the extent of housing need in local areas (Central Housing Advisory Committee, 1939). They were more active in these ways, and in pioneering new management techniques, than in housing production. As we have seen, the 1920s and 1930s were decades of expansion for public-sector and private speculative housebuilding and, latterly, for slum-clearance

Plate 7.1 A self-builder and his family in Sheldon, Birmingham in 1949. Although self-building and the voluntary sector generally did not receive much positive support from the post-war Labour government, the fifty participants in this scheme managed to obtain a loan from a building society and embarked on the work after a training period of six months. The fifty bungalows in the scheme were allocated in an order determined by ballot. If the little girl, who would now be in her late forties, should chance to see this, perhaps she might contact the author to say how well the scheme succeeded. Photograph: Topham Picture Source.

schemes. In 1935, however, the National Federation of Housing Societies (later Associations) was formed as a central organisation aided by government grant to assist and support individual societies. One of its first tasks was to undertake a survey of voluntary associations and their resources, from which over 170 organisations were recorded (Jones, 1985). New members were rapidly recruited and Parliament lobbied for increased support. As a result the Housing Act, 1936 provided for subsidies to associations, for access to the Public Works Loans Board and for associations to enter into agreements with local authorities.

In the years after the Second World War the voluntary sector had no clear role in the Bevan-inspired drive to increase housing supply. In the eyes of the Labour government it was local authorities who properly had statutory powers and a democratic mandate to carry out clearance and reconstruction schemes, and to whom therefore resources should be channelled. Both the voluntary sector and speculative housebuilders were regarded as not 'plannable instruments'. Nevertheless housing associations carried out schemes for specific needs such as providing housing for industrial workers funded partly by the employers, and they became increasingly involved in fostering self-build activities. They were perhaps most prominent in providing housing in which the growing number of elderly people could live independently – an activity which did have Bevan's positive support. Both new and existing trusts and associations built groups of flats and bungalows, sometimes with a resident warden, in association with church organisations and other voluntary bodies such as the Womens' Royal Voluntary Service and the British Legion. In this way they often complemented the activities of the local authorities, who more often built for 'general needs' in this period.

Very often, however, the 'general' did not include the growing number of Black and Asian immigrants who were settling in Britain. The racial tensions that built up towards the end of the 1950s, especially in London and Birmingham, were very often centred on housing issues because the residential qualification and prejudice alike worked to exclude immigrants from access to local authority housing. Organisations like Aggrey Housing Ltd in Leeds, the Birmingham Friendship Housing Association and the Nottingham Coloured People's Housing Society, by contrast, catered explicitly for them (Jones, 1985). This capability to identify and deal with the needs of minority groups was one of the considerations that led to a governmental rethink about the role that the voluntary sector could play in housing provision. In particular, minds were focused by an alarmist letter to *The Times Weekly Review* of 4 December 1958 from the secretary of the National Federation of Housing Associations, who warned: 'The danger of "doing nothing about it" is revealed by the growth of Nazi power in pre-war Germany.'

NEW ROLES AND POWERS AFTER 1961

It was in the early 1960s then that the voluntary sector began to play a more important part in the provision of rented housing. Under the Housing Act, 1961 loan capital was made available from the Ministry of Housing and Local Government for housing associations to develop housing for rent at levels which neither required subsidy nor produced a profit. The £25 million made available for England and Wales was taken up in the development of about 7000 new dwellings by the end of 1963. Under the Housing Act, 1964 a new government agency, the Housing Corporation, was set up to register associations, to provide funding and generally to oversee the operation of the voluntary movement and encourage the growth of new associations. The overall aim was to provide rented housing to be let at moderate 'cost' rents, perhaps especially to those at the beginnings of their careers who could not yet afford to buy and who tended to be moving around the country fostering the 'white-hot technological revolution' that, according to prime minister Harold Wilson, was about to transform the economy.

The development capital was to be in the form of mortgage loans coming normally two-thirds from private sources and one-third from the Housing Corporation. As interest rates rose, however, the rents were driven higher to cover costs, and this form of housing became less attractive compared to council renting, where the 'pooling' arrangements produced lower rents, and to purchasing, which benefited from the MITR. As a result, by the early 1970s pure 'cost rent' schemes were becoming less common than co-ownership schemes, where the resident acquired some equity interest and received tax benefits, or co-operative schemes, where a 'share' in the property or the common areas was purchased.

Voluntary-sector organisations were also becoming more involved in the improvement and conversion of existing older housing in inner urban areas. This often involved purchasing large single family houses and converting them into a number of flats which were then made available to those with special needs – for example the disabled or single-parent families. Special subsidies were made available for these purposes under the Housing Subsidies Act, 1967 and the Housing Act, 1969. Very often the problem was to find the funds to acquire property and pay for the conversion work some time in advance of the beginning of the flow of rentals. Associations looked to sponsors or donors while, at the national level, the charity Shelter raised and distributed funds for this purpose. Help was also made available by church bodies such as the British Churches Housing Trust and the Catholic Housing Aid Society. From 1969 too the Greater London Council set aside about £25 million per year as loan capital to assist voluntary housing agencies.

The Housing Act, 1974 provided a clearer legal definition of 'housing associations' and set up a new funding system which provided for 'HAGs' or Housing Association Grants. This took the projected annual income from any given scheme on the basis of 'fair rent' levels (see Chapter 6) and net of maintenance and management costs. It then capitalised this annual sum into the size of debt that it would service. The capital sum required for the scheme would be larger than this, perhaps typically four times larger, and it is this capital deficit that was made available as a HAG. Thus the finances of the scheme balance from its first year of operation, and as rents rise with inflation a surplus on the scheme begins to appear. For a period the associations were permitted to retain these surpluses but under the Housing Act, 1980 they had to be remitted through Grant Redemption Funds. Thus the amount of public subsidy in voluntary housing schemes is difficult to assess, since rents will have grown differently in different parts of the country, but it is certainly less than the original grant level. The HAG arrangements shifted the voluntary housing movement towards the public sector, at least in terms of the source of most of the funding. But it also had the effect of eroding the incentive to control costs and raise rents, since it was by nature a 'deficit-funding' arrangement.

CHANGES FOLLOWING THE HOUSING ACT, 1988

This combination of large-scale public-sector spending and lack of 'market discipline' in the way it was spent by largely autonomous local associations was predictably anathema to the Conservative administrations of the 1980s. Following lengthy discussions, the frame of operation of the voluntary housing movement was radically changed again under the Housing Act, 1988. The effect has been, for one commentator, a 're-privatisation' (Randolph, 1993). There has been a deregulation of the rents of new tenancies, so that associations themselves can determine rents, and a number of moves to attract private capital finance into the movement in place of public funds. The mixed-funding system introduced in April 1989 provided that only up to 75 per cent of the development cost of any scheme would be available as a HAG; the rest must be raised as loan finance in the private capital market. One clear aim of this is to limit public expenditure in view of the alarming increase in the Public Sector Borrowing Requirement. Another intention is that each scheme, and the asset base and general management capability of the borrowing association, should be subject to the scrutiny of market lenders. This introduces an element of 'market discipline' into the management of both the expenditure and the income side of the accounts.

These changes rested on the assumption that the private capital market would be interested in lending to what are often very small associations. Moreover they are organisations whose purpose in the political rhetoric, the provision of 'social housing', has not yielded conspicuously large returns in the past. Certainly there has been some hesitancy in the market (see European Capital Company Ltd, 1991), and the major financial institutions have not shown much interest. One helpful development has been the setting-up of The Housing Finance Corporation (THFC) in 1988. This is in essence a financial intermediary which seeks to match the needs of a large number of small borrowing associations and the lending preferences of much larger institutions. The THFC raises money in large 'issues' and then distributes it in the required 'packages' to borrowing associations. It has so far channelled several hundred million pounds by these means (The Housing Finance Corporation, 1993).

The effects of the 1988 Act on the voluntary sector have been assessed by various commentators (for example Burrows, 1989; Randolph, 1992; Stevens, 1993). One effect has been to raise the overall level of rents charged as older 'fair rents' are 'harmonised' with newer 'assured rents' – in a context where the externally set 'fair rents' themselves rose by an average of 18 per cent in 1990/1 compared to general inflation of 7.1 per cent (Stevens, 1993, 5). The overall effect was an increase of 81 per cent, compared to general inflation of 26 per cent, in the rents of new lettings over the period 1988–91 (Randolph, 1993, 45). Further increases are expected in the future as the ratio of HAG to private-loan finance decreases. The average proportion of development costs to be financed by the HAG component is expected to fall to 67 per cent in 1993, 60 per cent in 1994 and 55 per cent in 1995. This will mean a rapidly increasing exposure to the costs of debt servicing and to the vagaries of interest-rate changes in the private capital market. It seems inevitable that either output will fall, or the quality and space standards will fall, or rents will continue to rise sharply – or there will be some combination of these effects.

Much of the cost of the rent rises will fall on the Housing Benefit system, the cost of which is escalating alarmingly (Edmonds, 1992; Burrows, Phelps and Walentowicz, 1993). Already in many housing associations the majority of tenants are eligible for Housing Benefit. Thus, as in the case of local authority housing discussed in Chapter 6, publicly funded support is being shifted from stages 2 and 3 to stage 4 of the model in Figure 3.1. This shift from supply-side to demand-side support has been judged by many critics to be exactly the wrong thing to do. The grant of £750 million to the Housing Corporation, announced in the 1992 Autumn Statement, for the acquisition of 17,000 houses may also prove to be a mixed blessing (Birch, 1993). This appears to have been distributed between the Housing Corporation's nine regions more on the basis of indicators of over-supply of speculatively built private housing (for example the rate of fall of new house prices and the number of new houses unsold) than in terms of more acceptable indicators of housing need. Clearly this would serve to get some speculative house-builders off the recessionary hook. If this assessment is even partly correct, it justifies the growing sense that the Housing Corporation is in some ways 'the social housing sub-committee of the House Builders' Federation' – as it has been dubbed by some commentators. In all, the effects of these changes call increasingly into question the appropriateness of the voluntary sector as the main provider of good standard 'social housing' at 'affordable' rents.

The increase in output of new housing association units also failed to fulfil the promise of a 'development boom' in the years after the 1988 Act (Randolph, 1993, 42–4). This was due partly to a cash-flow crisis in the Housing Corporation (see Harrison, 1992, 27). The effect may well be a short-term one as the Corporation and associations familiarise themselves with the new arrangements and new possibilities of raising finance from the private capital market (European Capital Company Ltd, 1991). Certainly in 1991/2 associations registered with the Housing Corporation completed over 27,000 new units, which represents nearly four-fifths of 'public-sector' provision (an increase from less than 3 per cent in 1975/6 – see *Roof*, March and April 1992, 11). Approvals for future development were 11,000 over target, at 51,000. A total of over £428 million of private-sector lending had been secured (Housing Corporation, 1992). Nevertheless these rates of development are nowhere near the rates achieved in the buoyant periods of local authority building (see Chapter 6) and they are clearly not matching the increase in need for low-rent accommodation. There are also clear signs that associations are becoming more competitive with each other for the funding available. In the view of one commentator the winners are likely to be 'the leaner, fitter and more entrepreneurial organisations' (quoted in Harrison, 1992, 30). It remains to be seen whether these institutional characteristics will be the ones to help those in greatest housing need.

WHAT GETS BUILT – AND REBUILT?

Those elements in the built environment which have been built or renovated by the voluntary housing movement are probably more varied in age, scale and design than those produced by the processes discussed in the previous three chapters – although virtually all of them are produced to provide residential accommodation. Because at no time has the movement been the main provider of social housing, but instead has tended to focus on certain categories of need, it has never built large spreads of readily recognisable and homogeneous development. Rather, its products hitherto have been small-scale, well scattered throughout urban and rural areas and often built with a sensitive eye to the needs of the intended users. A fair amount of the total development has been in the form of renovation or redevelopment and for this reason solutions designed by an architect have been more common than 'off the peg' designs. In all, the movement is characterised by diversity – not only of type of institution (see Baker, 1976, chapter 3) but by type of housing produced.

The housing built by the charitable trusts, some dating from a century or more ago, is still very much in use. The Joseph Rowntree Trust financed the building of the village of New Earswick near York which used 'garden city' principles and other design innovations. The Peabody Trust manages about 12,000 dwellings and 'Peabody Buildings' are to be found in a number of cities. Over 23,000 people are housed in the Bournville Village Trust properties, still to a high standard. Similarly the Guinness Trust used the provisions of the Housing Act, 1974 to refurbish a number of tenements it had built in London in the 1890s. These originally had a tap on each landing and free baths for the use of the residents – excellent amenities for the day. Numerous other trusts continue to produce and manage housing. For example the Samuel Lewis Housing Trust, endowed by a merchant banker of that name in 1901, is working to provide housing for the Bengali community in Spitalfields, east London.

Another category of voluntary organisation was the locally based housing associations that grew up between the wars. As an example the Bethnal Green and East London Housing Association started from Christian origins in 1926, attracted gifts, issued loan stock and built and rehabilitated flats in Hackney and Poplar. It grew after the war by amalgamation, attracted more funding from the Greater London Council, the boroughs and central government under the 1974 Act, and currently owns 500 homes in east London. Similarly in 1928 the Liverpool Improved Houses Association was founded and soon benefited from the handing-over of several hundred properties by the Marquis of Salisbury. In the late 1960s the Association formed a partnership with the local authority and Shelter to improve run-down inner-city houses. Now, as Merseyside Improved Houses, it operates across the region and has a new-build programme including sheltered units for the elderly and special housing for ex-psychiatric patients and the physically handicapped. It currently manages over 12,000 homes and in addition has set up separate organisations to foster job creation and environmental improvement. A final example is the North Eastern Housing Association. This was set up in 1935 by the government itself to assist in the depressed areas of Tyneside and Durham. It received Exchequer grants and payments from the Special Areas Fund. By 1939 it had built 8000 homes and by 1975 it owned over 18,000. In 1980 it merged with the North Housing Association, its present title, and currently it has a stock of 21,000 homes.

Coming closer to the present day, organisations in the voluntary housing sector are active in building a bewildering variety of schemes which are usefully summarised elsewhere (for example Jones, 1985; and Royal Institute of British Architects/National Federation of Housing Associations, undated). To give a flavour, Homes for Elderly Vegetarians Ltd provides sheltered housing in a number of areas for elderly people who unsurprisingly 'must be 100 per cent vegetarian'. Following Housing Corporation encouragement there were, by 1976, eighty-six housing associations providing accommodation specially designed for the physically handicapped. Prompted by MIND, the National Association for Mental Health's 'Home from Hospital' campaign in 1976, a number of housing associations focused on the needs of patients discharged from mental hospitals – a group much swollen in numbers as a result of the 'care in the community' policies of the 1980s. Similarly NACRO, the National Association for the Care and Resettlement of Offenders, has co-operated with organisations in the voluntary sector in the provision of hostels and other schemes for people discharged from prison. A more recent problem is the accommodation needs of those discharged from the armed services as a result of defence expenditure cuts (Brimacombe, 1991). Another problem raised by the dramatic change in world politics since the late 1980s has been the housing needs of the influx of refugees from war-torn areas of East and Central Europe.

The voluntary housing sector can often respond to needs such as these with a flexibility and speed that is currently beyond both the bureaucracy and the financial capability of cash-starved local authorities. On these grounds, and certainly in terms of the dedication of those working within it, it may be seen as a positive feature of our housing system. But it can be seen in another way – as a means of providing good housing standards for a proportion of those in need but in so doing as partly concealing the extent of the problem and failing to deal with the acute problem of homelessness. As at 1991/2, for example, only 17 per cent of the lettings of the twenty largest housing associations were used to house the statutorily homeless (*Roof*, March/April 1993, 11). The sector can be seen as a safety net that stretches only halfway along the tightrope. As such it will be most effective when acting as a complement to local authority housing provision, as it has effectively in the past, not as a substitute for it.

There is now a serious concern that government policy is forcing a change in this role – perhaps largely because while to neo-liberals the notion of public-sector provision is anathema, voluntary-sector housing activity is 'consistent with the anti-collectivist principles of the Conservative government' (Back and Hamnett, 1985). By pushing the voluntary sector into the provision of 'general needs' social housing, expecting it to take over a greater share of the problem of coping with homelessness and forcing a reduction in building standards by reducing the central grant, the government is giving the sector new tasks while decreasing available resources (Page, 1993). There is an increasing worry within the voluntary sector itself that associations will be induced to build larger estates, perhaps on poorer and more isolated sites and to lower design and construction standards. These estates are likely to house an increasing proportion of people on very low incomes and with a range of social needs for whom many associations do not yet have well-developed management procedures. This may lead to estates falling into a process of decline, as happened with large council estates that were poorly designed, maintained and managed.

There is also a concern that as more capital finance comes from private-sector borrowing rather than state grant the lending institutions will look more closely at the asset position of the borrowing association than at the housing role it is seeking to fulfil. This is likely to mean that many smaller associations which currently house minority groups in need, including ethnic minority groups, may be taken over by larger associations. The diverse nature of the sector and its current excellence in serving small areas of need may thus be undermined. Many practitioners in the voluntary sector are worried at the implications of this scenario and would wish to return to their more traditional role of working in partnership with local authorities, not as commercially funded replacements for them. As one recent commentator put it: 'what will be the future nature of housing associations: servants of their communities and consumers or self-serving, second class property developers?' (Cope, 1990, 293).

SELF-HELP HOUSING

Self-help housing falls in the voluntary housing sector since it is neither statutorily required nor developed for profit. It may be defined as housing 'where the first occupants arrange for the building of their own dwelling and, in various ways, participate in its production' (Duncan and Rowe, 1992, 1). It takes one of a number of forms – either the future occupant acts simply as promoter, acquiring the site and contracting a builder, or she/he acts as both promoter and constructor; or some other agency (such as a local authority) may provide the site and finance and the self-builders may then build their homes and rent them. In practice these various categories may shade into each other. In any of the cases the finance might come from a variety of sources or it might be from the accumulated savings of the occupant. The self-builder who promotes the scheme, uses her/his own money and subsequently maintains the property provides a rare case of all five stages of the model shown as Figure 3.1 being carried out by the same person.

Over the period 1980–9 self-help housing accounted for only about 6 per cent of British output. It has therefore attracted comparatively little attention in British housing literature, which has, unhelpfully in some ways, been obsessed with tenure issues (as pointed out by Barlow and Duncan, 1988). But in this respect, as in so many other ways, Britain is out of step with European practice. Over the same period self-help housing provided 60 per cent of the output in Belgium, more than 50 per cent in West Germany, Austria, Italy and France, and more than 40 per cent in Norway, Finland and Ireland. In almost all these cases the proportion of single-family houses produced by self-help was much higher.

Producing housing by means of self-help almost always means a saving in cost. Thus a house of similar quality to that available from 'the market' can be obtained for less money, or a better house can be obtained for the same money. A number of studies have shown that self-help housing provides higher levels of satisfaction than industry-produced housing (Duncan and Rowe, 1992, 15). The cost-saving arises partly from the savings in labour costs either simply in the promotion process or in both promotion and building – depending on whether it is self-build or not. Other savings arise from not having to pay builders' overheads and profits. Cost savings in the range 20–30 per cent for self-building and 5–15 per cent for self-promoting are reported, and figures from Atlantic Canada, West Germany and Norway confirm this general picture (see Harms, 1992; Rosnes, 1987; and Rowe, 1989). Self-help housing seems not to provide a general answer to the shortage of accommodation for lower-income people or vulnerable groups, since most studies show that it is carried out mostly by middle-income groups, often in the 30–45 age group. On the other hand those that do house themselves this way have a very secure tenure. There is less chance of suffering repossession, because the debt on the property is lower in relation to the value than for non-self-help housing, and in a fair proportion of cases no debt was incurred anyway.

There are a number of barriers to undertaking self-help housing. Clearly certain categories of people such as the elderly and disabled find greater difficulty in self-building, but can nevertheless self-promote their own home on equal terms with others. Various estimates have been made of the order of time involved in self-building. Several studies show a 6–9-month period using holidays, weekends and summer evenings, and a German study shows 2000–2500 hours spread over up to two years. If a simplified building technology is used, such as the Segal or Claxton systems in Britain or one of the kit systems widely available in Sweden, France and Canada, lack of previous construction experience has not been found to be a barrier to self-building. In fact in countries where one or two generations back most of the population were largely rural there may well be surviving or inherited building skills and traditions. When members of the British Housing Advisory Programme in Bulgaria (see Chapter 10) raised the possibility of self-build housing as one of the means by which the previous highly centralised housing system could be reformed they quickly learned that many Bulgarians had a very strong latent capacity for self-building which forty years of socialism had failed to extinguish.

Housing policy, planning regimes, land-supply agencies and financial institutions can be supportive to self-help housing by making sites, planning consent and finance easily available. This is the case in most European countries and in Atlantic Canada. Or, as is the case in Britain, the planning system can be less helpful, the building industry more hostile, the lending institutions more suspicious and the land market 'tighter' and more prone to speculation. All these conditions help to explain why Britain lags behind most European countries in this form of housing provision. As Duncan and Rowe (1992) point out, self-help housing in the rest of Europe is not mainly associated with backwardness or peripherality. It is a major element in the expansion of Frankfurt, Paris and Stockholm. Policies and attitudes to support this form of housing provision in Britain seem long overdue.

AN EXAMPLE OF A VOLUNTARY-SECTOR AGENCY – BRIGHTON HOUSING TRUST

Brighton Housing Trust, which began operations in 1970, is a housing association registered with the Housing Corporation, and a registered charity. As at October 1992 there was a management committee of fifteen people, the central administration of the Trust was carried out by a full-time director, a deputy director and a staff of four, and the total staff employed numbered seventy-two. The main office is in an accessible part of central Brighton adjacent to the main Brighton Housing Department office.

Financial assistance and other income are provided by a very large number of organisations, including not only the Housing Corporation, the Department of the Environment, the Department of Social Security, Brighton, Hove and Eastbourne Borough Councils, East Sussex County Council, the Regional Health Authority and Shelter, but also local businesses, charitable trusts and foundations, schools, churches, media organisations and many members of the public. The turnover of the Trust in a typical year is in excess of £6 million (including both capital and revenue transactions) and the value of the assets is somewhat more than this.

The Trust has built up a range of services to assist those in housing need in a wide variety of ways. As a housing association the Trust has a stock of over 150 accommodation units, most of them suitable for the needs of single people or one-parent families. Some of the flats are designed to be used by people in wheelchairs. In fact one impending development of sixty homes is 'barrier free' – all the units and the spaces between are wheelchair accessible. Access to the accommodation is either via the Housing Departments of Brighton and Hove councils or via the Trust's own waiting list. The housing management service offered is more intensive than that normally available from a local authority because many people housed by the association have additional problems, perhaps stemming from homelessness, poverty or ill-health, and they may require additional help to live independent lives 'in the community'. As part of this policy tenants are encouraged to participate in decisions about the maintenance and management of the property they live in. In addition to this accommodation the Trust has a stock of rooms in shared houses for people in need of specific forms of support, perhaps because they are recovering from alcohol or chemical abuse or have mental-health problems.

Some of the rooms are in Resettlement Hostels and are particularly for single people who have been without a permanent home for what may have been a long period of time. Here, apart from accommodation in single rooms with shared facilities, people are offered help and support in the process of gaining the self-confidence and skills needed to live 'in the community'. The Recovery Project, which is designed to help those recovering from alcohol/chemical dependency, has a total of sixteen beds for those at varying stages of recovery. The approach is based on the World Health Organisation philosophy which sees dependency as an illness in its own right. The treatment offered includes encouragement to abstinence, one-to-one counselling, group work and educational presentations. The staff team includes a qualified counsellor and several people at the postgraduate stage of clinical-psychology and social-work training. Apart from the 'in-house' treatment there is a non-residential service available which includes advice, counselling and after-care.

The Trust also operates a day centre, which is regarded as central to its total range of activities and which can call on the combined expertise of all staff.

The centre is located in a former church which was extensively renovated in 1988. It is aimed at the needs of those sleeping rough or in bed and breakfast or lodgings accommodation. The centre has an 'open-door' policy and is used by about 150 people daily. A free breakfast and cheap lunch are provided, together with showers and laundry facilities. In addition there are recreational facilities such as table-tennis, pool, games, an art room, books and a literacy class. People requiring healthcare are referred to medical services as necessary, and help is offered to make claims for benefit entitlements. Apart from the 'in-house' staff there are two outreach staff who offer support for those made extremely vulnerable by lack of any proper accommodation. These also act as a crisis team, giving advice to those sleeping rough, or taking people to hospital if necessary. The centre, which is easily reached on foot from the centre of the town, is well known on the 'grapevine' to those requiring its services.

Finally the Trust operates a Housing Advice Centre in both Brighton and Eastbourne. This offers free and confidential housing advice either by appointment or on a drop-in basis. The topics covered include how to find accommodation, housing costs and welfare benefits, and how to deal with the housing implications of relationship breakdowns. The purely legal advice is dealt with by the Legal Project. This is staffed by a full-time legal staff including three solicitors and five legal caseworkers. It offers a full range of free legal advice, including representation in court and emergency services. The areas covered include security of tenure, mortgage arrears and possession proceedings, disrepair problems, illegal eviction and harassment, rent problems, problems with leases, duties of local authorities, and entitlement to welfare benefits. The project has achieved successes in, for example, obtaining injunctions against illegal evictions, arranging repayments of mortgage arrears over a period, getting rents reduced by Rent Assessment Committees and obliging local authorities to accept responsibility for homeless families.

Brighton Housing Trust provides an excellent example of the integrated way in which housing and other services can be delivered by a voluntary sector agency. It is not formally democratically accountable (and is therefore in the NDA sector in Figure 3.1) but it works very closely with the appropriate departments of the local authority, primarily with the Housing Department, and is very sensitive to the views of its tenants. In addition its activities are monitored by the Housing Corporation. It is run by an expert and dedicated staff who can respond creatively to needs and make decisions faster than would be possible via the committee structure and timetable of the local authority. It has also expanded across 'departmental' boundaries to serve areas of need, such as alcohol/chemical dependency, which while not necessarily rooted in housing are very likely to have a housing dimension. It enjoys the goodwill of many right across the political spectrum and can thus continue its activities despite changes in the colour of the local authority. It is a model of power-sharing by means other than the ballot box.

CASE STUDY – THE GOLF DRIVE DEVELOPMENT IN BRIGHTON

This is a development of sixteen one-bedroom flats arranged in pairs. It is designed as homes for single men and women who have either previously been street homeless or have come from bed and breakfast or hostel accommodation. The development was promoted by Brighton Housing Trust (BHT) and it opened early in 1992.

The site was sold to BHT by Brighton Borough Council at the full market price applicable for 'social housing'. The entire funding was in this case obtained from the Housing Corporation. The grant element covered 58.10 per cent of the costs, and the remainder was borrowed from the Corporation at full market rate. The grant proportion depends on the particular mix of each development, for example whether 'special needs' units are included, and on its geographical location. This percentage was below the Housing Corporation average (72 per cent) for the year in question. The construction was carried out under a fixed-price contract because since the Housing Act, 1988 cost over-runs have to be borne by the promoting association. The development has a number of special design features. Each flat is entirely self-contained and has separate access with no common areas. This helps to

Plate 7.2 Some of the one-bedroom housing units in the Golf Drive development promoted by Brighton Housing Trust in a quiet residential area a mile or so north of the centre of Brighton. Each of the upper and lower units is fully self-contained and common spaces have been minimised. Photograph by Peter Ambrose.

keep management costs and therefore rents down. The insulation is of better specification than building regulations require since the tenants may not be able to afford high heating bills.

The nomination rights for tenancies rest 50 per cent with the local authority (the legal minimum for joint schemes) and 50 per cent with BHT. Rents in 1993 were £45.29 per week with a weekly charge of about £1 for services. Tenants are assisted if necessary by the advisory services of the Trust in claiming Housing Benefit and any other benefits for which they are eligible. Since the tenants are likely to have very little furniture and few household goods on arrival, both carpets and kitchen appliances are provided, and the Trust advertises for other furniture and so on that may be required. The scheme, in common with others developed by BHT, benefits from a relatively intensive management service with an average of one housing-management staff member per forty-eight tenancies.

When the scheme was at the planning stage, a meeting was held to which local residents were invited, since there was a concern that there might be some resistance to the housing of homeless men and women in the area. In fact the proposals gained the approval of the residents and there have been no subsequent complaints. The tenants have planted some trees and taken good care of the immediate surroundings and it is clear that if people are placed in a good environment they take care to keep it good.

In all the scheme provides well-designed, well-managed and appropriate housing for a group who would otherwise be either street homeless or housed in expensive and inappropriate emergency accommodation.

Part III

POWER

New ideologies and their effects

The processes generating the built environment which have been discussed in the previous five chapters form, collectively, an extremely important element in the general economy. They consume considerable private-investment and public-finance support, employ a substantial proportion of the labour force, have an impact on the balance of payments and produce the infrastructure and accommodation for all forms of human activity. Analysis of the way they work, and the output they produce, can be meaningfully carried out only if proper account is taken of the predominant overall political belief system in which they are operating and of the policies generated by that belief system.

The two chapters that form this part of the book deal with the striking changes that have occurred in prevailing political and social ideologies since the early 1970s, the policies that have flowed from them, and some of the effects of those policies.

Chapter 8 traces the origins and growth of 'neo-liberal' ideas and the ways in which they have become integrated into economic and social policy, especially under the Thatcher administrations from 1979 to 1990.

Chapter 9 examines some of the impacts of this new hegemonic politics on the evolution of the built environment in terms of the agencies and institutions that influence development, the reduction in the degree of democratic accountability, the changed pattern of promotion and output, and the growing body of evidence suggesting that the output has been an inappropriate response both to market demand and to social need.

8

THE DOMINANT 'NEO-LIBERAL' IDEOLOGIES OF THE 1980s/1990s

In the depths of the Second World War there was overwhelming popular support in Britain, especially in the armed forces, for the emerging concept of a post-war 'welfare state'. This was encapsulated in an edition of the popular weekly *Picture Post* (4 January 1941) which included a series of short articles by some very well-known names setting out the way things would be when the fascist powers were defeated. Healthcare, education, housing and planning, and various other forms of social provision were to be made available on egalitarian principles and largely by publicly funded and democratically managed agencies. The landslide Labour victory of July 1945, and the defeat of the great wartime leader Winston Churchill, demonstrated the popular enthusiasm for such policies as a means of achieving a better post-war society. As several commentators have pointed out, the reference point for the 1945 election was not so much the Battle of El Alamein in 1942 as the hardships of the 1930s.

Some years into the Labour administration the sociologist Marshall argued that while a degree of inequality was an inevitable outcome of the capitalist ownership of the means of production, and that capital's search for profitability would continue to some extent to be a source of this inequality, the situation would be partially redressed by the 'citizenship' rights inherent in the newly legislated 'welfare state' (Marshall, 1950). A market-orientated development of the economy would generate sufficient wealth to finance the programmes and policies that would provide insurance against poverty. Systematic income differentials would still remain but reforms in insurance, welfare benefits and education would lead to a greater 'equality of status'. The 'mixed economy', while not achieving all social-policy aims, would work in a more benign fashion than that of the less interventionary pre-war days.

Recently Wilson has argued that the post-war health and welfare legislation did indeed produce a political order which provided social stability for several decades after 1945. In his view the 'post-war consensus' had four essential elements: 'Commitments to full employment, comprehensive state social welfare provision and foreign policy were the broadly agreed ends. The principal means was a mixed economy managed according to principles derived from the theory of John Maynard Keynes' (Wilson, 1992, 21). Certainly for nearly three decades after the war there was relatively strong cross-party support for the social policies which emerged following the Beveridge Report of November 1942 (Beveridge, 1942), the Education Act, 1944 and the post-war legislation which established the National Health Service and brought about the reform of the historic Poor Laws. Similarly, governments of both parties sought to limit unemployment and manage cyclical fluctuations in economic activity by increasing or decreasing public spending and other forms of demand management. Conservative governments were content to leave largely undisturbed the nationalisation measures of the post-war Attlee government, while the Labour Party did not seriously dispute the need for sensible control on social expenditure; nor did they materially undermine the financial and political power of the major financial institutions. In all:

> On social issues, Conservatives against much dogma accepted the goals of full employment, state welfare, and the principle of equal opportunity in education. Labour (in practice) tolerated substantial unemployment, accepted class-based private sectors of provision of health, education and welfare, and imposed limits to the standard of living of workers, through statutory wage restraint.
>
> (Wilson, 1992, 22)

THE END OF THE POST-WAR CONSENSUS?

There has been very considerable debate about whether the period since the early to mid-1970s constitutes a break with this 'post-war consensus' in British politics. At about this time these consensual policies were widely perceived to be failing to provide effective management of the British economy in an increasingly competitive world. The evidence began to refute the expectations. Keynesian theory suggested that unemployment and inflation should alternate. The former could be reduced by a dose of reflation and increased public spending to regenerate the economy and stimulate activity. This would lead, via nearly full employment and the enhanced leverage power of organised labour, to increased wages, production costs and prices, and thus to inflation. This in turn could be controlled by decreased government spending and/or an increase in interest rates. There were two major assumptions here; one was that economists understood the processes involved, and the other was that national governments control the decisive forces at work. Both began to look increasingly shaky. Keynesian theory was to some extent falsified as inflation and unemployment began to occur concurrently in the early 1970s. Simultaneously the growth of the new industrial powers and the loss of colonial markets had undermined Britain's competitive position (see, for example, Gamble, 1990). Finally the developing world-scale investment strategies of trans-national financial and manufacturing organisations moved them increasingly beyond the regulatory power of national governments.

In terms of social policy the post-war consensus was called increasingly into question for another set of reasons. First it did not seem to have achieved its main aim of abolishing poverty and bringing about a more equal society. A celebrated report (Abel-Smith and Townsend, 1965) showed that the proportion of the population below an acceptably defined poverty level was much the same in the 1960s as it had been in the 1890s. The number of people dependent on non-contributory benefits to raise their income above the 'poverty line' rose from 1 million in 1948 to nearly 3 million in the late 1970s. Finally a rigorous study produced in the mid-1970s concluded that: 'there is no clear indication in recent years of a trend towards the elimination of poverty, defined in terms of the official Supplementary Benefit scale. In these terms, poverty has remained a problem of considerable magnitude throughout the last twenty years' (Atkinson, 1975, 198). The second main problem with the post-1945 welfare policies was their cost. In an era when the British economy was slowly slipping down the world league, public spending on welfare broadly defined (including social security, personal social services, healthcare, education and housing) rose from 16 per cent of GNP in 1951 to 29 per cent in 1975. This clearly led to acute problems in raising revenue to finance the programmes. If the increased expenditure were to be raised from increased taxes on those in employment the inevitable response would be increased wage claims to protect the level of take-home pay. This would make British goods even less competitive. If company profits were taxed more heavily, this too would lead to increased prices at the point of sale and to further inflation. Taxing wealth other than at the point of inheritance has never been popular in British politics and the main form in which it used to exist (rates on property) was never likely to yield revenue on the scale required. All this militated against the Beveridge-cum-Keynes vision of achieving an egalitarian society and a stable economy largely by means of substantial public-expenditure programmes. As Deakin later observed: 'public expenditure, regarded throughout most of the post-war period as a benign instrument for securing the goals of social policy and the final guarantor against the return of economic depressions had changed sides: from being an essential part of the solution it had now become part of the problem' (Deakin, 1987, 72).

THE NEW RIGHT AND ITS KEY IDEAS

At this juncture, in the early 1970s, a successor paradigm to that of the post-war welfare state began to emerge in the form of a set of ideas collectively known as 'neo-liberalism' or the 'New Right'. As many have pointed out, the New Right was not new at all but had as one of its roots an individualism and economic liberalism that can be traced back to the Middle Ages (see, for example, Green, 1987, chapter 1). Certainly some of its central ideas, for example concerning a least-interventionary state and a naturally occurring social hierarchy, long predate ideas about universal welfare provision. Partly the set of ideas derives from Adam Smith. One of his concerns was that the interests of the consumer should be protected against the rapacity of manufacturers and merchants (Smith, 1776). While he felt that governments could play some part in this process, he believed more firmly in the benign effects of the 'invisible hand' of self-interest on the part of entrepreneurs. Self-interested motivations and actions, when 'channelled' by sensible laws to promote and ensure competition, would in fact lead to outcomes that benefited all.

This idea that universal benefit will flow from the operation of the 'invisible hand' of self-interest among the economically powerful is clearly contentious and has been subject to considerable examination, notably by Hayek (for example Hayek, 1967). It runs counter to the more generally held view that societal organisation is a zero-sum game – that is to say that gains for some must entail losses for others. At a more colloquial level it confounds the Yorkshireman's dictum that there are no free lunches. Many would argue that any genuine spreading of benefit from increases in wealth generation must depend upon state-managed 'trickle-down' effects. These, the argument runs, take the form of positively discriminatory measures in the fields of taxation and means-related welfare payments. From this viewpoint it is difficult to to justify a political strategy, such as that of the 1980s and early 1990s, which invokes the 'invisible hand' argument, systematically withdraws state-managed redistribution policies and then appears to deny the rising levels of inequality which have so clearly resulted.

A second crucial element in New Right thinking derives from other aspects of the multi-faceted work of Hayek, especially his arguments about a 'natural order' in society and the practical need for people to have established customs, codes and laws for guidance rather than to have constantly to rework the functional relationships between means and ends. To live our lives on the latter basis would, he argues, be impossibly burdensome – if only because we know so little about the way in which means and ends relate. He identifies two kinds of social order (see, for example, Hayek, 1973). One is the result of some imposed human design and is in that sense a *made* order. Such a society is organised for some predetermined purpose. It can be at the level of a medieval village, the armed forces or the Third Reich. But at whatever scale it must be amenable to constant oversight by the person or persons 'in charge'. Hayek's thinking here reflected his deep concern about what might result from the 'made' order of Hitler's highly centralised and totalitarian Third Reich – not surprisingly since he had been born in Austria and had witnessed the events of the 1930s at close hand. His worries centred around strong state planning, since in his view the role of the state was to create conditions favourable to progress rather than positively to plan it. He articulated the case in *The Road to Serfdom* (Hayek, 1944). The book was, ironically, completed at the London School of Economics – the institution powerfully shaped by Beveridge, one of the chief architects of Britain's post-war social planning.

It may be added that in Hayek's 'made' order there is an inherent danger that the totalitarian leader will work to achieve in the public mind a correspondence between his or her aims and those of 'the society' or 'the people'. A variety of means, perhaps including terror, may be used to impose a centrally directed set of aims and values, while propaganda will seek to convince the citizens that these imposed aims and values are collectively held. In other words people will be encouraged to internalise and own the prescribed values. Some may feel that we have seen a mild version of this in 1980s Britain with the endlessly repeated rhetoric about the virtues of self-interest, competition, nationalism, and so on. In this situation there is a danger of concluding that it is 'society' that has produced certain legislation and espoused certain values when in fact what has happened reflects the projection of self-interest, whether ideological or financial, on the part of a narrow segment of interests. There is a connection here with a point made in Chapter 1 about the politically disabling acceptance that 'society' or perhaps 'the market', rather than definable groups of interests, produces observable outcomes – for example housing shortages or the over-production of offices.

Hayek's other type of social order is 'spontaneous' and it has something in common with Adam Smith's 'invisible hand'. Here there is no grand design, although those that make up the order will all have designs for their own lives. Order emerges, and is maintained, by the widespread acceptance of laws, habits, codes of morals, and established institutions with predictable behaviour, including governments. But in a passage that exactly sums up much of the policy of the Thatcher years, and the growth of the notion of the 'enabling state', Hayek argues that the role of government is 'not to produce any particular services or products to be consumed by citizens, but rather to see that the mechanism which regulates the production of those goods and services is kept in working order' (Hayek, 1973, 47). There are several reasons for this, in Hayek's view. Governments simply do not know enough, either of the aims and motivations of citizens or of the way millions of people interact. As a result they cannot reliably predict all the consequences of intervention and regulation – in fact centrally imposed policies

151

are bound to have widespread unintended consequences. One could hardly quarrel with this – one sees evidence of it daily.

Another reason for Hayek's distaste for too much government intervention in the workings of society is that it is frequently aimed at the production of more 'social justice'. Certainly most would agree that this was the general intention of most of the British post-war welfare state legislation. But Hayek's view, and that of a succession of subsequent 'New Right' theorists, notably Sir Keith Joseph, is that the pursuit of social justice is self-defeating and inevitably undermines individual freedom and initiative. They argue that while the actions of an individual can be called just or unjust, the concept of 'justice' is not an appropriate way to approach the analysis of societal phenomena such as differences in wage-rates or housing conditions. In their view to hold to such a belief is to support the notion that some arbitrary power in the form of governments, rather than market mechanisms, should establish and impose norms concerning the value of an individual's labour to others, or the distribution of material possessions. Hayek's view of the matter was uncompromising – this is not an appropriate aim of government. Thus the term 'social justice' should be abolished from the language (Hayek, 1976). In fact one of the great benefits of societies organised on 'free market' lines is precisely that myriads of judgments do *not* have to be made by government about issues such as wages and incomes. Such a line of argument, which implies that markets rather than elected governments should crucially determine material outcomes, raises questions about the rationale for democratic practice. The logical end product of the argument is startling: we should apparently exercise power more in our role as consumers in a market or as individual vendors of our labour than by the act of voting. The power we exert via the ballot box produces agencies, central and local governments, whose exercise of power in many important spheres of decision-making is undesirable.

Many New Rightists, then, share a set of beliefs. 'Freedom' is ensured not by governmental action but by the interplay of producers and consumers in markets, both for goods and for labour. This economic freedom is a route to political freedom (Friedman, 1982, especially chapter 1) – although many commentators from Adam Smith onward have recognised the need for *some* degree of governmental action to ensure 'free competition' and the protection of consumers. The prices ruling in the market, moreover, should be as far as possible 'undistorted' by public subsidy. Too much governmental action, whether in the form of subsidy support or in terms of regulation of the monetary value placed on commodities such as labour and particular goods, undermines freedom and choice. It renders the citizen subject to the 'arbitrary' will of others. It gives a distorted rather than a 'true' picture of prices to the consumer, and it discriminates capriciously between the worth of an individual's labour based on some scale of values determined by a public authority. It is better for society to be regulated by 'the rule of laws' in the sense of established custom and practice, rather than the 'rule of men' (*sic*) in the sense of governmental edict. It is pointless to search for 'social justice' in the form of equality of outcomes, although many New Rightists do believe that there should be equality of opportunities. To seek for equality of outcomes is to deny the undeniable – that people have widely differing capabilities; in a memorable phrase: 'if you make people equal at breakfast time there will be a pecking order by lunchtime'.

SOME PROBLEMS WITH THE NEW RIGHT

This brief review has not of course done justice to the sophistication and complexity of thinkers such as Adam Smith, Hayek and the subsequent theorists of the New Right such as Scruton (their ideas are accessibly reviewed in, for example, Bosanquet, 1983; and Green, 1987). But the analysis and the prescriptions can be seriously challenged at many points. Modern production systems offer increasingly complex goods and services matched by ever more sophisticated means of persuading people to aspire to things which they did not know they needed, and increasingly diverse ways of enabling them to borrow money in order to buy them. In this situation it is difficult to see how anything short of comprehensive government legislation – for example concerning production standards, methods of advertising and the management of credit – can provide adequate consumer protection. One remains sceptical of the protective power of the 'invisible hand' of producers' and merchants' self-interest. This often seems to land people in massive debt and surrounded by goods which often have a finite life programmed in and which do not provide commensurate degrees of satisfaction. It also seems to be instrumental in generating spin-off problems as diverse as widespread obesity, high rates of lung cancer and damage to the ozone layer.

One can criticise too the over-narrow definition of 'political power'. Much of the New Right literature reads as if directive influences and limitations on freedom emanate only from governments, while 'the free market' ensures that none of us is subject to the arbitrary will of others. But obviously very considerable power over the range of goods and services available, and thus patterns of consumption, is exercised by the large-scale producers in the market – especially those who control the research and development and manufacturing processes that produce new items for mass consumption. Visibly these often act as cartels. In addition only the naive would believe that the pricing process for commodities such as petrol and compact discs is genuinely competitive. Thus control over the workings of the market is exercised by a variety of powerful organisations of which only one is government. The key difference between these organisations is that government, under democratic arrangements, is removable by the exercise of the vote. The others are not. There are also the negative freedoms – the 'freedoms from ...' – to be considered. These are often most powerfully safeguarded by elected agencies of the state. Even a centralised and bureaucratic socialist state such as the pre-1989 GDR delivered more freedom from unemployment, eviction, malnutrition and hypothermia than systems organised on 'free-market' lines – however repressively it behaved in other ways.

Hayek's dismissal of the concept of 'social justice' is likewise contestable. One's instinctive test of justice or fairness seems applicable to all acts, whether those between individuals or those carried out by a private organisation or a government. A decision by a company to sack all employees with brown eyes or an Act of Parliament laying down that all those suffering from diabetes are, *ipso facto*, to be evicted from their homes would surely be universally regarded as unjust, as *socially* unjust. Similarly legislation specifying the collection of taxation on some basis widely held to be unfair (for example the Poll Tax) is another clear example of injustice not between individuals but at a societal scale. It follows from these that justice must have a social dimension. If legislation, and the conduct of organisations or governments, can be regarded as socially just or socially unjust, then there must be merit in the search for social justice, whatever Hayek may say on the matter. This is not to say that policies need go so far as to include tight governmental regulation of all prices and wage-rates. The search may well be for arrangements that help to ensure generally acceptable degrees of fairness between producers and consumers in markets, whether for goods, services or labour. In other words one is looking for a level playing-field – although of course there will be endless disagreement about what precise degree of declivity constitutes a slope.

Another problem arises concerning the New Right's reliance on the concept of 'free market prices'. The belief in the efficacy of the 'free-market' as the most effective and moral way to further economic and social development requires that prices at the point of sale should be 'undistorted' by government action in the form of subsidy and regulation. This is so that the price structure visible at the point of purchase should present the consumer with a 'true' set of signals about the relative value of commodities and thus allow him/her 'genuine choice' in spending preferences. Again the antithesis to these arrangements was provided by the GDR, where social, political and even artistic judgments, and the need to reduce the consumption of imported goods, had pervasive effects on the centrally planned structure of retail prices and other living costs. Relatively simple and perfectly acceptable cameras, hi-fis and consumer durables were cheap in relation to incomes. More complex versions of the same goods were prohibitively expensive or unobtainable. Classical long-playing records were 12 marks, 'pop' records 16 marks. Books were extremely cheap and so were seats at the opera. Rents were normally between 4 per cent and 10 per cent of incomes. The state-imposed pricing structure was itself an arm of policy and must, over the longer term, have affected social behaviour and cultural preferences.

It is a central thrust of the 'reforms' all over 'East Europe', insisted upon with great force by neo-liberal 'Western' agencies such as the International Monetary Fund and the World Bank, that these arrangements should be ended as soon as possible. Prices should be deregulated and should float free in the market place without any 'distortion' from state action. Economic and pricing systems should become, in their terms, 'clean'.

But there is a gaping hole in the argument here. No price is 'free' in this sense. The collective and collusive actions of the powerful players in the market themselves affect prices at the point of sale. More fundamentally, the price of any product at the point of retail sale normally reflects production cost plus one or more 'mark-ups', plus perhaps the 'distorting' effect of tax. This composite cost in turn incorporates research and development effort that may have been partly carried out by public agencies. It also depends partly upon the movement of production inputs and the finished goods over a road or rail network funded by public money, the use of energy inputs derived partly from public spending and, of course, the use of labour educated and kept healthy largely at public expense. Thus a whole range of costs involved in the production process have been 'externalised' and collectivised across the society as a whole. The calculations necessary to arrive at a truly 'undistorted' production cost, disregarding the effects of all this past and present state investment, simply cannot be done. Thus the 'free prices' based on these costs do not reflect purely privately financed effort. The New Right argument that 'free market prices' convey to the consumer a 'true' message about the relative 'value' of the various goods and services on offer, and thus enhance individual freedom by making consumer choice more 'genuine', fails to incorporate the complexity of the production process.

The argument that 'free prices' incorporate some element of morality is even more puzzling. State governments surely carry some moral responsibil ity for the long-term development of their society and the long-term good of the planet – at least politicians continually make speeches which would have us believe this to be so. Yet many goods which are widely sold because in narrow 'market' terms they are highly profitable, for example aerosols and nitrates, appear highly damaging to the planet. Other goods, for example cigarettes, are highly damaging to people. Yet other goods and services, for example decent housing and healthcare, are demonstrably good for people. Surely the moral path for governments is to recognise these truths and intervene in price structures, not ignore them. Morality seems to dictate that the prices of goods and services known to be environmentally and socially damaging should be regulated at so high a level that nobody buys them, while access to goods and services that are accepted to be beneficial should be actively supported by reducing their price. Most governments try to produce these effects by their indirect taxation and subsidy policies. So at least the *principle* of intervening in the price structure at the point of sale in the interests of the longer-term collective benefit of the society or the planet has been conceded. It is clearly an intervention based on some notion of collective morality – and very sensible too. But it makes little sense for the New Right to argue, in a context of growing concern for the environment and the longer-term health of society, that it is ultimately moral that all prices and rents in the market place should be 'free' and 'undistorted'.

A final point of criticism concerns the notion that a 'rule of laws' is to be preferred to a 'rule of men' (*sic*) as the organising legal framework for the just society. As Hayek argues (Hayek, 1973), laws can be directions or commands or they can be 'rules of just conduct'. The distinction is useful but not clear-cut, because no doubt all governments would argue that they legislate the former because they *are* the latter. Few outside government will be satisfied with this. But the present criticism is on slightly different grounds. To assume that some set of 'rules of law' exists that is independent of the 'rules of men' is an ahistorical fallacy and it disables clear analysis. The 'rules of law' are simply the 'rules of men' inherited from some previous time period. The fallacious assumption is like some superficial appeal to 'historical forces' when seeking to explain current phenomena. There *are* no 'historical forces' – there are simply specific explanatory factors which have operated for some length of time and have thus attracted the adjective 'historical'. In fact when either 'historical factors' or 'rules of law' are invoked one should beware. Both have their origins in times past when both power structures and social and political conditions may have been very different. They may be neither good explanation nor good law for the present day.

THE POLITICAL IMPLEMENTATION OF NEW RIGHT IDEAS

New Right, or 'neo-liberal' ideas have been shown to have a long history. In fact they can claim to be the historical norm when one considers the thousands of years of human economic interaction that preceded the advent of the interventionary and welfarist modern state only a century or so ago. Since that time neo-liberal ideas have been advanced into the political arena by agencies such as the Economic League, which was formed in 1920 to advocate the benefits of capitalism and warn against subversion from political undesirables. A similar organisation called 'Aims for Industry' was founded in 1942. Soon after the war, in 1947, Hayek was instrumental in founding the Mont Pelerin Society, so called because it met on this minor eminence above Lake Lucerne. Its aim was to involve both industrialists and economic theorists in discussions aimed at furthering the long-term development of free-market institutions at a time when, in Britain at least, the 'command economy' adopted during the war was still being to some extent im-plemented by the Labour government.

Ten years later, having read some of Hayek's work, the chicken magnate

Antony Fisher was active with others in setting up the Institute for Economic Affairs, which has since published a great deal of material mostly advocating free-market solutions to economic problems. Subsequently, in 1974, the Centre for Policy Studies was founded by Sir Keith Joseph and Margaret Thatcher following the election defeats of the Heath government in that year and the feeling among many Conservatives that the mould of their party's thinking, which included regulatory 'mistakes' such as prices and incomes policies, needed to be decisively recast. The CPS has therefore been more closely associated with the Conservative Party than the IEA. Finally the Adam Smith Institute was founded in 1977, by which time Mrs Thatcher had been the leader of the Conservative Party for over two years, sterling had been 'rescued' by the International Monetary Fund (with a number of policy-formation 'strings' attached), a new post-Keynesian orthodoxy had swept the world, and the Heath years were clearly long gone. All these institutions have been active in producing a stream of publications advocating the main thrust of the new economic orthodoxies. Several of these were highly influential in setting the Conservative agenda in the run-up to the 1979 General Election (for example Bacon and Eltis, 1976; and Howe *et al.*, 1977).

What have been these main thrusts? The economic policies of the Thatcher administrations, although they have naturally wavered around in response to political realities, have been held to with a reasonable degree of consistency. They have been reviewed by a number of commentators (for example Riddell, 1991; Johnson, 1991; and Wilson, 1992). Chief among them has been the drive to reduce and control inflation. In general terms this has been successful – although as it happens the Retail Price Index of annual inflation was slightly higher when Mrs Thatcher left 10 Downing Street than when she first entered it. But this is hardly fair. Considering the longer term, the 1970–9 average annual rate of inflation was 12.5 per cent; that of 1980–9 was 7.4 per cent (Johnson, 1991, 72). Nevertheless it is a matter for debate how much of this effect was due to Britain's internal economic policy and how much to the fall in the world price of various key commodities that occurred in the early 1980s.

The second main aim was to reduce the amount of money borrowed by the state to finance its annual expenditure programmes, that is to say the Public Sector Borrowing Requirement (PSBR). Since a simultaneous aim was to reduce the overall amount of taxation, it clearly followed that public expenditure must be reduced. To quote Howe *et al.*:

> it is so important to reduce the share of the nation's wealth consumed by the state – by central and local government and those agencies and authorities which spend the taxpayer's money but produce nothing.... Our intention is to allow state spending and revenue a significantly smaller percentage slice of the nation's output and income each year....

157

This will be in contrast with Labour's recent panic cuts, which fell too heavily on capital rather than current spending and did great damage to the construction industry.

(Howe *et al.*, 1977)

The assertion that central and local governments somehow 'consume' money and that the roads, railways, power grids, housing, hospitals, schools, universities, etc. that they produce amount to 'nothing' is manifest nonsense. Looking back over the 1980s, the construction industry may also feel that the claim that public capital expenditure would be protected has not been borne out. The depression in almost all sections of this industry has been long and severe partly as a result of the severe contraction in state capital programmes.

For any given level of expenditure, a shortfall in funds raised via the PSBR and aggregate taxation can be made good by money received from the sale of public assets. This is termed either 'privatisation' or 'selling the family silver', depending on one's political position. Paradoxically, it was a Conservative prime minister (Harold Macmillan) who coined the latter phrase, and Mrs Thatcher who regarded the sales as a procedure to restore 'public' property to the family. Many families will wonder, if it was public already, why they had to pay to regain it. This is exactly what many people are wondering in East Europe at the moment as the state sells them 'their' flat. The confusion is rooted in a difference of opinion about whether or not 'the state' can be regarded as 'society' – i.e. some kind of collective family. For Mrs Thatcher, of course, it could not.

Moving from the imponderable to the measurable, something like £60 billion was realised in the 1980s by the privatisation of, for example, utilities and the sale of land and buildings (Johnson, 1991, 172). The true market value must have been very considerably higher, since a high proportion of these sales, in all asset categories, were at sub-market or discounted prices. In fact the difference between the total sum realised and the true market value, if it could be computed, would represent in one sense the cost of the policy. The annual proceeds increased considerably through the 1980s and peaked at £13 billion in 1988–9 (Johnson, 1991, Table 32). The PSBR requirement for that year was negative (–£14.5 billion), partly as a result. The money received from public-asset sales rose to be a very significant source of total state revenue; it was, for example, 7.2 per cent of general government expenditure in 1988–9 as against only 1.3 per cent at the beginning of the decade. The problem for the government as the 1990s unfold is that the PSBR is threatening to rise to unprecedented levels. This requirement stems partly from the ideologically based drive to reduce taxation and partly from the increased cost of other fall-outs of New Right policy such as high unemployment. If, as is feared, the PSBR rises to £40 billion or £50 billion annually, it could not possibly be offset by the sale of public assets since beyond a certain point there would not be enough left to sell. Future reductions in the requirement may well have to come from yet more stringent controls on state expenditure or from increased taxation. Neither will be easy to achieve.

The guiding economic theory of the Thatcher administrations, especially in the first few years, was that of monetarism. Critics are inclined to see this guideline as a piece of dogma rather than a theory. The overall notion is that inflation can be reduced and controlled by reducing and controlling the money supply and thus aggregate demand. This raises two main problems. The first is to define precisely what 'the money supply' means, and the second is to devise a means to 'control' it. In a complex modern economy, money takes a wide variety of forms, from notes and coins to various forms of credit. The total supply of money at any one time, and the aggregate demand that it produces, depends not only on the volume of notes and coins in circulation, but also on the behaviour of a vast number of individual credit-granting institutions, most of them in the private sector. Following the various deregulation measures of the early and mid-1980s, designed to induce greater competition between financial institutions and increase 'consumer choice', there seemed no obvious way that the government could control the volume of credit emanating from the private sector.

THE ECONOMIC AND SOCIAL EFFECTS

There is now an extensive literature on the economic and social effects of the post-1979 Conservative administrations (see, for example, Riddell, 1991; Andrews and Jacobs, 1990; Johnson, 1991; Cloke, 1992; and Wilson, 1992). Ministers and other apologists for the government claim that inflation has been reduced, output, productivity and employment increased, investment increased, taxes reduced and the power of organised labour brought under appropriate control. The results of the imposition of monetarism on the Chilean economy as from 1973 have also been widely discussed (see Wilson, 1992, 77–9). There have been a number of economic benefits, including more rapid repayment of external debt, at the cost of grievous social consequences such as mass unemployment and a marked increase in income and wealth inequalities. Most of the same effects, in a muted form, are evident in Britain and they are becoming characteristic of the former socialist economies of East and Central Europe.

The British economic record has been subjected to sustained criticism, and Wilson, a less than friendly observer, disputes virtually all the government's claims (Wilson, 1992, 81–98). He makes the points that the level of production achieved in 1974 was not regained until 1988, the gains in productivity were achieved more by rising unemployment than rising output, and in 1990 Gross Domestic Product (GDP) per capita in Britain stood lowest but two (Italy and Greece) in an OECD comparative study. Levels of investment as a percentage of GDP were much lower in Britain

than in the main competitor countries and by 1986 investment in manufacturing industry had fallen to only 78 per cent of its 1979 level. In fact in
the decade up to 1989 investment in new plant and machinery in manufacturing rose by just 8 per cent (the corresponding figure for 'financial services'
was 300 per cent) and this was probably a factor in causing Britain to become
a net importer of manufactured goods for the first time ever in 1983. The
trade balance in manufactured goods changed from a £5 billion surplus in
1979 to a £20 billion deficit in 1989 – an effect attributed by Wilson in part
to the importation of 'consumer goods purchased with the proceeds of
substantial tax cuts and a credit boom fuelled by uncontrolled borrowing
against inflated house values' (Wilson, 1992, 84). If even partly correct, this
demonstrates a further fallibility in the notion that 'monetarist' policies will
control spending and inflation. In addition to all these, Wilson points out that
Britain's share of world trade declined from 8.2 per cent by value in 1979 to
6.8 per cent in 1988; that by September 1990 the balance of trade on
'invisibles' was zero and that there was a massive deficit on tourism and
travel; that British companies consistently invested less both in research and
development and in education and training than their main overseas
competitors; and that British government support for basic research had
fallen dramatically. This seems to be a fairly comprehensive set of adverse
indicators.

Broadly similar conclusions were reached by Johnson, a less hostile
commentator, who remarked:

> By the end of Mrs Thatcher's reign it looked as if the U.K. had learned
> nothing from its own recent economic history and seemed condemned
> to repeat it. The 1990s show opened just as the 1980s show had opened:
> the gloomy recession act took up most of the time before the first
> interval; seat prices had doubled; and everyone was 25 per cent better
> off than they had been ten years before but fearful of finding they had
> no job when they came out.
>
> (Johnson, 1991, 255)

Not everybody, of course, could afford the price of a seat, because not
everybody was 25 per cent better off. The failure of the New Right policies
of the 1980s to reduce social inequalities to what many would regard as
acceptable levels has been documented in Chapter 2 of this book. Various
commentators on the Thatcher years have pointed to the ways in which
social policy has produced, sometimes clearly intentionally, greater social
inequalities on a wider societal scale. Recent work (for example Andrews and
Jacobs, 1990; and Wilson, 1992, chapter 16) has shown conclusively that both
in broad policy intentions and in the fine print of legislation the 'reforms' in
the fields of social security and health have had clear objectives. These
include a reduction in social expenditure, a real-terms reduction in the value

of many benefits, a pervasive trend towards means testing, the exclusion of increasing numbers of people from benefit, and the encouragement of private or voluntary organisations, rather than those with a democratic mandate, as the means of providing health and social care.

The overall results of these policies are hardly contestable. A report from the DSS (Department of Social Security, 1992) estimated that in 1988/9 about 12 million people were living below half-average income, a reasonable definition of poverty; in 1979 the corresponding figure had been 5 million. About 25 per cent of all children were living below this line in 1988/9 (10 per cent in 1979). The poorest one-tenth of the population had suffered a 6 per cent cut in real income in contrast to the rising prosperity of those on above-average incomes. The share of total income received by the poorest half of the population fell from 32 per cent in 1979 to 28 per cent in 1988/9. The minister responsible was quoted as saying that the adverse effects on the poorest groups were partly attributable to rising unemployment and to dependence on self-employment (*Guardian*, 16 July 1992). He felt able to conclude: 'Overall it is clear that the message is good.' Opposition spokes-people pointed out that the report provided conclusive evidence that the 'trickle-down' effect simply did not exist.

SUMMARY

This chapter has sought to identify the massive changes that have occurred in the implicit and explicit orthodoxies that dominate British politics, primarily in the period since the early 1970s. It has thus provided a context for the following chapter, which is concerned with the consequent changes that have occurred in the processes shaping the built environment. The philosophical foundations and the social and economic assumptions for the new orthodoxies have been discussed. The ways in which they gained first credence and then dominance in the political debate have been traced and so, in outline, have some of their main economic and social effects.

The form of politics that did as much as anything to produce these changes has been termed an 'elective dictatorship'. One of its central characteristics was an explicit rejection of the search for consensus. This can best be judged from a remark made by its chief architect: 'To me, consensus seems to be the process of abandoning all beliefs, principles, values and policies.... It is the process of avoiding the very issues that have got to be solved merely to get people to come to an agreement on the way ahead' (Mrs Thatcher, quoted in Kaldor, 1983, 89). Given the deeply troubled history of the twentieth century, it is chilling to reflect that the search for societal harmony and agreement can be reduced by a British prime minister to a 'merely'. Demagogues through the ages have used such arguments to justify their 'solutions'. Mrs Thatcher achieved and held power by the very democratic means she clearly despises. If such views, enunciated by someone in supreme power, became the common currency of British governments, it would make even more grievous the plight of the poor and bode ill for the future political well-being of the nation.

9

THE IMPACT OF 'NEO-LIBERAL' POLICIES ON THE BUILT ENVIRONMENT

This chapter will seek to show some of the ways in which the neo-liberal ideologies and politics discussed in the previous chapter have had effects on the environment-generating processes dealt with in Chapters 3–7 and on their output. This is an ambitious project and it is undertaken in the recognition that there are no simple cause–effect relationships in matters so complex. Changes in the built environment, while primarily politically determined, do occur partly in response to influences which are not overtly political – for example as a result of changes in building or transport technology. But even here one is on shifting sands, because technological and political processes are intimately bound up in each other. For example it would be extremely naive to believe that the decline of investment in the railway system and the growth of spending on roads was purely a reflection of technological change in these two forms of transport infrastructure. It would be equally innocent to ascribe the 1970–90 changes in land use in London's Docklands entirely to the new facilities required by containerisation.

Inevitably this chapter is highly selective in the material it covers, and this opens it up to the charge of bias. It may well be the case that in some respects the greater influence of 'market forces' which has stemmed from the implementation of neo-liberal politics and policies has had demonstrably positive effects. Some of the buildings produced in the 1980s were excellent by almost anyone's standards – for example the *Financial Times* building. Similarly the planning regime of the 1980s and early 1990s, although it has depended to an extent on the varying attitudes of a number of Secretaries of State, has on the whole taken a firm conservationist line in the face of pressures to cover the rural landscape with out-of-town shopping centres and so-called 'new towns'. In these respects the influence of one hundred or so Tory MPs from the leafier areas of the home counties has not been negligible. But the main thrust of this chapter is that in most human terms, and even by a number of strictly economic criteria, the overall impact of New Rightist policies on the built environment has been adverse. Those who disagree can write their own books to refute the case.

This chapter will focus on events and changes at various stages of the model shown as Figure 3.1. It will include an analysis of the changing NDA:DA balance in both the promotion (stage 1) and production (stage 3) of new built environment. It will then consider changes in housing-rent levels – a crucial arbitrator of access (stage 4) – and present data to show that rent rises stemming partly from changing allocative criteria (see Figure 1.1) have had grievous effects for many of those least able to compete in the housing market. Subsequently it will deal briefly with an example of a 'market-led' institution set up specifically to deal with a particular redevelopment situation (London Docklands). It will summarise the reactions of several people who have experienced at first hand the adverse economic and social effects of the policies. It should be noted that in previous chapters (notably Chapters 6 and 7) attention has been drawn to a further characteristic of neo-liberal strategy – that of shifting public support for built-environmental provision from stages 2 and 3 of the chain to stage 4 – roughly from supply side to demand side.

CHANGES IN THE NDA:DA RATIO OF PROMOTION IN THE 1980s

It was argued in Chapter 3 that different stages in the process of providing and renewing the built environment carry different degrees of social sensitivity. For example events at the promotion, allocation and subsequent management stages may well have a more directly felt effect on people's needs and life-chances than those at the financing and construction stages. It is therefore in the former stages that user participation in the determination of events may be most desirable. How, then, has the total provision process changed over the period of New Right political dominance since the late 1970s? How, at each stage, did the balance of NDA to DA influence and funding change? What consequent shifts have occurred in the nature of motivations and accountability? Has 'democracy', as defined, gained or lost ground?

It would take more than this chapter, or the entire book, to provide a comprehensive answer to these questions. The production of each major form of built environment depends upon a specific set of organisations and agencies and specific funding arrangements. We could take motorways and other roads as an example. These are promoted by central and local government departments with statutory responsibility for road construction and maintenance. Investment finance is still largely from public revenue sources. Construction and subsequent maintenance and repair are normally

164

carried out by private-sector civil-engineering contractors. Allocation is primarily by market mechanisms – anyone who can afford a vehicle, and possibly in the future a toll, can use them. With the exception of the very few 'unadopted' roads, the promoting organisations maintain a high degree of control at all stages – except over who uses the roads (stage 4 in Figure 3.1). One could work through the production of, for example, hospitals and recreational facilities in a similar way. But for present purposes attention will focus on the forms of built environment which together occupy the largest proportion of land area in urban areas and which are of greatest social sensitivity – housing and commercial schemes such as offices and shopping centres.

The institutions and agencies involved in the provision chain for these types of built environment, and the ways in which they interact, have been fully discussed elsewhere (see, for example, Balchin, Kieve and Bull, 1988; Ball, 1988; Cadman and Austin-Crowe, 1991; Goodchild and Munton, 1985; Jackson, 1973; Rose, 1985; and Short et al., 1986). But so far there has been little discussion examining the relationships between first the changing macro-political context, second the changing 'weight' of NDA and DA promotion, and third the resultant changes in the pattern of orders placed for new construction (stage 1 in Figure 3.1). This discussion can be developed by considering the data presented in Table 9.1:

Table 9.1 New orders obtained by contractors (£ millions at 1985 prices)

	New housing		Other new work			All
	Public	Private	Public	Private		
				Industrial	Commercial	
1978	2113	4108	4075	2389	2687	15,372
1979	1512	3805	3803	2389	2466	13,975
1980	882	2739	3410	1923	2553	11,507
1981	789	2648	4206	1718	3090	12,451
1982	1115	3647	3872	1455	3077	13,166
1983	1050	4686	4564	1723	3108	15,131
1984	911	4294	4386	2425	3772	15,788
1985	734	4555	3877	2149	4028	15,343
1986	746	4896	4240	2088	4624	16,594
1987	808	5202	4232	3463	6054	19,759
1988	712	5431	4264	2654	7499	20,560
1989	643	4046	4836	2630	7727	19,882
1990	498	2851	4150	3054	6423	16,976

Source: HMSO, *Housing and Construction Statistics*, Table 1.1

This gives a picture of the changing pattern of promotional activity by two broad groups of clients, 'public' and 'private'. Although, as the discussion in Chapter 3 made clear, these two categories are not regarded as exactly coterminous with the NDA and DA division used in Figure 3.1, the equivalence will be regarded as sufficient for present purposes. In 1978, the last full year of the previous Labour government, the proportion of all construction contracts placed by DA clients was about 40 per cent – £6,188 million in a total of £15,372 million. In 1981, in the depths of a construction recession and before neo-liberal policies begun to be fully felt, the proportion was still 40 per cent. As the recovery took place, private orders rose much more sharply than public and the DA proportion fell to 34 per cent in 1984, 30 per cent in 1986 and 27 per cent in 1990. Since, as has been argued, the pattern of promotion conditions much of what happens further 'down-stream' in the model these are very significant changes. It is reasonable to hypothesise that as a result the promotion of new built environment for profit-seeking purposes (roughly those dealt with in Chapters 4 and 5) has become more important in relation to promotion driven primarily by 'social' considerations (forms of development covered in Chapters 6 and 7). Has this in fact occurred?

CHANGES IN THE PATTERN OF OUTPUT

In 1978 housing orders constituted about 40 per cent of all new work – £6,221 million out of £15,372 million. By 1984 this proportion had fallen to about 33 per cent and by the end of the period it was less than 20 per cent. In real terms the value of all housing orders in 1990 was only 54 per cent of what it had been in 1978 – a very considerable fall over a short period. Within the housing sector the proportion of orders placed by DA promoters fell from 34 per cent in 1978 to 18 per cent in 1984 and to less than 15 per cent in 1990. Two effects are therefore discernible over the decade or so. First, far less housing was being promoted both in absolute terms and as a proportion of all construction effort. Second, of this diminishing share a fast-diminishing proportion was being placed by DA promoters. Clearly, in relation to housing at least, the NDA/DA balance of effort and influence at the promotion stage increased sharply in the period covered by the data.

In the 'Other new work' category, similar trends are evident. In 1978, DA agencies placed about 45 per cent of all orders. This fell to 41 per cent in 1984 and about 30 per cent in 1990. Over the decade or so DA 'other new work' fluctuated considerably but showed slight overall growth. Reference to the same issue of *Housing and Construction Statistics* (Table 1.10) will show that the major share of this growth related to road construction. NDA orders for 'industrial' work, primarily factories and warehouses, showed a deep trough in the early 1980s but recovered by the end of the decade to levels somewhat above those of the late 1970s. The really impressive growth of orders for construction has taken place in the NDA 'commercial' category. This means primarily the construction of retail centres, offices, hotels and so on. Orders for these stood at £2,466 million at the beginning of the Thatcher years (at 1985 prices), rose sharply in the early years of the decade and stood at £7,727 million in 1989 – a real-terms increase of 313 per cent over the decade – before falling back somewhat as the recession bit deeper. These figures confirm the supply-side volatility previously identified in Chapter 4.

Unpacking the data a little, the value of construction work for NDA promoters in each of the main types of development rose in nominal terms as follows over the period 1980–90:

	%
Housing	+229
All industrial	+187
All commercial	+428
Offices	+522
Shops	+334
Entertainment	+359
All output	+276

Source: HMSO, Housing and Construction Statistics, Table 1.10

It would require considerably more effort to collate and present similar information on the changing NDA:DA ratio at stages 2–5 of Figure 3.1, but it is safe to conclude that the same order of shifts in the ratio has occurred at all stages. Public spending, especially on capital-investment programmes, has been increasingly constricted through the period, and since 'public' authorities rarely provide the investment for orders placed by 'private' clients it is reasonable to conclude that the investment shift (stage 2) is at least as striking as that at stage 1. At the construction stage (stage 3), indicative evidence of the shift is available from the statistical source already quoted. Table 1.8 of Housing and Construction Statistics details 'Direct Labour Output' or construction work carried out by local authority building departments. While, as Table 1 above shows, the value of all construction orders rose in real terms by over 47 per cent in the 1980–90 period, work carried out by Direct Labour Departments fell by 24 per cent (at 1985 prices) over the same period. New housing work was only 12 per cent of what it had been. Even the amount of housing repair and maintenance, one of the main forms of Direct Labour work, fell in real terms. Given that the total amount of housing maintenance and repair carried out by all contractors rose in real terms by 32 per cent over the decade (Table 1.7 of the same source) the Direct Labour departments have clearly lost some share. This provides evidence of the changing NDA:DA ratio at stage 5. At the allocation stage (stage 4) the same shifts have clearly occurred. Fewer local authority houses have been built, fewer lettings are available for allocation on need-related bases and very

167

few purely public-sector commercial developments have been carried out. In total far more built environment is being allocated by 'market' criteria and far less by 'social' criteria.

What conclusions can be drawn? First, that events at the promotion stage have become strikingly less democratically accountable over the New Rightist 1980s. More commissioning of new development has been carried out by profit-seeking agencies and less by the 'public' sector. The same tendency is evident at all later stages of the production process. Second, based on the motivational mainsprings in each sector, more development has been driven by capital-accumulative motives ('Assessment of market' in Figure 3.1) and less by 'social' criteria ('Assessment of need'). The figures presented are a rough and ready reflection of this. Housing production has lagged far behind the output of 'commercial' forms of development, especially offices, shops and entertainment facilities. And an increasing proportion of housing output has been allocated on market criteria anyway. The third conclusion is that, no matter what policies the planning and housing authorities may write into their strategy statements concerning low-cost housing and industrial development as a priority, the actual pattern of new built environment produced is moving in the opposite direction. The obvious development choices for capital-accumulative promoters and investors to make, based on their best understanding of the market, are for office, retail and certain forms of leisure developments. At least the choice *was* fairly obvious up to the late 1980s, when the recession really set in. It is not so clear, now that signs reading 'Lease for sale' and 'Offices to let – only £1 per square foot' are festooned along many high streets in the country. Canary Wharf presents a problem too.

INCREASES IN THE COST OF ACCESS TO HOUSING

So far as rented housing is concerned, and as long as there is shortage in relation to need, the most socially sensitive housing issues are those to do with allocation (stage 4). The criteria in operation here determine who can gain access to accommodation, on what entry qualifications and at what personal or household cost. Before presenting data on the movement of housing-rent levels, and on a number of standard indicators of housing stress, it is as well to consider the alternative bases on which housing rents can be arrived at. The rent-setting variants, and the assumptions that underpin them, include the following:

Variant 1

> *assumption* – housing should be treated as a 'free good' because 'shelter' is considered to be in a category with other social essentials such

as education and healthcare and they should be largely or entirely collectively financed

rent policy – rents are set at a small or even zero percentage of income so as effectively to remove the user's labour-market position as a determinant of housing access; thus the larger part of both the capital costs and the current costs of running the stock comes from central subsidy; there is no relationship between the rent charged and the capital and running costs of the unit occupied

Variant 2

assumption – rents should cover the true current annual cost of the housing units occupied; initially this will require some subsidy if levels are to be remotely 'affordable', but if the unit is part of a public stock which has been built up over time some of the costs may have become 'historic' as a result of inflation, and current rents can be averaged out to cover the current costs of the whole stock owned

rent policy – rents structures are worked out annually in such a way that the total current revenue from the stock owned equals the current gross outgoings set out in a separate dedicated account (the Housing Revenue Account in the British case); the rents for individual units can be graded in any convenient way so long as the total for the 'pooled' stock equals total annual outgoings (subsidy may or may not be channelled to the account)

Variant 3

assumption – rents should be set purely in relation to the local supply/demand situation for accommodation of this kind; in other words the unit is a 'market good' which will find its price locally like any other

rent policy – rents do not reflect any social judgment and they have no logical relationship to the original construction cost of the unit; they simply reflect what the market locally will bear in current prices

Variant 4

assumption – the net rent gained on any given unit should reflect a competitive rate of return on the capital sum represented by the property when compared to competing forms of investment with equal risk and management costs; this will provide an adequate incentive for financial sources to invest in providing rented housing

rent policy – rents will be set so that the net return on the current value of the property is x per cent where x is an acceptably competitive rate of return; targeted rent allowances will probably need to be paid to enable the market to provide this level of rental return

Variant 5

assumption – people should pay roughly y per cent of their incomes in rent

rent policy –   rents are set or legislated so that people on roughly average incomes pay y per cent as rent; those on lower incomes may receive rent allowances; there is no logical connection between the rental flow and either the capital and revenue costs or any expected rate of return on the value of the property

Reasonable arguments can be made both for and against most of these – although there appears to be little sense in variant 5. Most of the socialist countries used versions of variant 1. They made assumptions that decent housing should be within the reach of all, regardless of income level. Many would applaud the underlying social intention, and apart from that the strategy of largely eliminating the cost of this particular non-substitutable good from household expenditures removes one source of pressure for wage demands. But because of shortcomings in the development and implementation of policy the socialist countries of Europe typically finished up with a heavily insensitive management bureaucracy, masses of sterile high-rise blocks, energy inefficiencies, no effective element of tenant participation in management, and a rental stream which was not sufficient to cover even basic management and maintenance costs.

For much of the history of British public-sector housing some version of variant 2 has applied. As shown in Chapter 6, the early subsidy regimes went some way to closing the gap between the true cost of providing the housing, and the revenue that could be derived from giving access to it to people with a wide spectrum of incomes – although not for many years to the very poorest. Theoretically the 1919 Addison arrangements could have done this, and they can be regarded as a 'deep-subsidy' version of variant 2. Local authorities could decide how much housing to build and what rents to charge. The costs not covered by the rents could be recouped from the Treasury 'open-ended' subsidy. It is not surprising that the arrangement hardly survived the short-lived political conditions that produced it. The rents charged averaged between 9s (45p) and 10s (50p) per week but sometimes were under 5s (25p) (Bowley, 1945, 25). These were all roughly comparable with the controlled rents in the private rented sector and no doubt generally offered a far better standard of accommodation.

The 1923 and 1924 arrangements in Britain represented a 'shallower-subsidy' version of variant 2. The Treasury contributed a flat-rate subsidy per house for a given period, and the local authority contributed one-half of this on similar terms. The rents were supposed to come out as similar to existing controlled rents. But clearly it might not work this way if costs got out of line with the limited subsidy offered. Rents would then be the variable factor and would inevitably rise. For this reason centrally imposed rules about rent levels had necessarily to be left vague. As the level of subsidy fell under successive regimes in the inter-war period, so the shortfall in the accounts had to be made good by increasing the rents of the housing occupied. Alternatively, or in combination, there could be an increase in the rate fund contribution (the RFC) which formed a mandatory element in the 1919 Act and in most housing finance legislation between 1924 and 1956. Naturally the precise combination in which these two means of covering a rents shortfall were used has often been a matter of lively political debate. Average council-house rents remained relatively low and stable between the wars, typically varying from under 6s (30p) in Leeds, Reading and Canterbury to over 9s (45p) in Blackpool, Brighton and East Ham (see Bowley, 1945, 115). The level depended partly on the building history of the authority: whether it had built mostly in low-cost or high-cost periods, under which specific subsidy regimes, to what extent an RFC was used and, no doubt, to what specific standards it built.

It should be noted that, until the 1972 Act, in Britain council rents remained relatively low in relation to the current costs of new additional stock. This was because as the stock 'matures' the current impact of loans taken out in the past falls away with the effects of inflation – gently if inflation is low, and sharply if it is high. For example the Brighton stock might still contain a house built in 1924 at a cost of £200 with a loan taken out at 4 or 5 per cent over a forty-year period. By 1960 inflation would have had the effect of reducing the loan-servicing costs to very low levels, in current money, and by 1965 the loan was paid off and these costs had fallen to nil. Yet the property still had a rental life lasting into the future and the rent charged for it (say £1500 per year in the 1980s) could be averaged with rents charged for more recently built properties to produce a 'profile' of rents lowered overall by this 'historic pooling' process. In other words the rents charged for the stock as a whole did not have to cover what would be the *current* loan-servicing costs of the total money borrowed to build it – as they have to do in variant 4 above.

The 1972 Act introduced the new notion of gradually moving council rents up to some level ('fair rents') set in relation to rents in the private sector – where a competitive rate of return on the current value of the capital represented by the unit *is* expected. Council rents were to be set first, at annually rising levels, and then subsidy needs calculated. The strategy was for the rising rents to 'pass through' variant 3 (market rents) and hope to reach the ideologically correct ground of a current 'market rate of return' on the publicly owned stock as set out as variant 4. The problem for housing investors, including local authorities and housing associations and their private-sector funders, is that in many areas low income levels and high unemployment mean that variant 3 rents (what the market will bear) may be well below variant 4 (rents providing an acceptable return on the capital value). The response to this problem by neo-liberal governments of the 1980s and early 1990s has been to increase user subsidy, currently Housing Benefit. The intention is to square the circle so that 'market rents' supported by user-targeted subsidy *will* in fact produce a rate of return that will attract sufficient new investment to the private rented sector. This will

171

ultimately obviate the need for any public rented stock – a form of housing with nothing to commend it in the New Right's ideological framework. This strategy is working up to a point but it is very expensive in terms of Housing Benefit, which currently consumes an increasing proportion of total housing support.

The 1972 strategy was exactly in line with the neo-liberal politics which, as we have seen in the previous chapter, came first to prominence and then to power a number of years later. But the problem was to ensure that local authorities *really did* raise rents at the rate required by central government in its underlying drive to let more housing costs fall on the current users of the properties. So the strategies of government in the 1980s have been rather different. The Housing Act, 1980 returned to setting subsidy first then letting rents 'fall out' of the financial calculations. The subsidy system was based on the amount paid in the previous year, updated by the Secretary of State's assumptions about upward movements on management and maintenance expenditure. Clearly upward rent movements could be forced on local authorities by the degree of miserliness built into the Secretary of State's assumptions. Governments of the 1980s have never been short on miserliness, so far as the public sector is concerned, and the 1980s have seen council-rent rises far in excess of general inflation. Another form of protection for rents was removed in the later 1980s when Housing Revenue Accounts were 'ring-fenced' and subsidisation from the rates became more difficult to achieve.

The overall effects of these policies has been council-rent rises considerably in excess of general inflation, as Table 9.2. shows:

Table 9.2 Increases in council rents 1979–92

Year	Average weekly unrebated council rents (£)
1979	6.40
1980	7.71
1981	11.43
1982	13.50
1983	14.00
1984	14.71
1985	15.59
1986	16.41
1987	17.24
1988	18.88
1989	20.76
1990	24.00
1991	27.49
1992	30.58

Sources: *Housing and Construction Statistics*; and CIPFA, *Housing Revenue Account Statistics*

The average rise in rents in all authorities in England between 1991/2 and 1992/3 is expected to be 11 per cent (*Roof*, July/August 1992, 11), compared to a general inflation rate of 3–4 per cent. Rises in the 'top twenty' authorities varied from 49 per cent (the London Borough of Camden) down to 22 per cent (Cheltenham).

The overall effect of reducing the amount of support for the financing and production stages (stages 2 and 3 in Figure 3.1) has been to drive rents above the level of affordability for many low-income or unemployed people, who then need to depend on Housing Benefit and similar support. Currently about 60 per cent of all public-sector tenants depend on benefits of these kinds. Similarly in the voluntary sector rents are rising sharply. The Housing Act, 1988 provided for housing associations to set their own rents for new lettings, having regard to 'affordability', and to decide whether and how to harmonise these rents with existing rents. At present three or four alternative bases for rent setting are being considered (see Stevens, 1993, 12). Mostly they derive from variants 2 and 4 above. In fact one basis is the Capital Values system, where the rent is set to reflect how much the vacant dwelling would fetch if sold on the open market. The claim that voluntary-sector housing is 'social housing' makes no sense in this case. The only certainty is that the general level of voluntary-sector rents will rise more quickly in the future than it has in the recent past.

Housing finance, it should be noted, is a complex subject and in the last decade or so there has been a bewildering sequence of changes in finance arrangements and the ways in which subsidy is calculated (well discussed by a number of authorities, for example Gibb and Munro, 1991; Malpass and Warburton, 1993; and more briefly in Malpass, 1992). The situation has been complicated still further by the 'right to buy' legislation, the use to which capital receipts may be put, the changing rules for raising capital for new schemes and renovation, and the effects of all this on Housing Revenue Accounts. The only reasonably safe assertions are that recent neo-liberal governments have sought ultimately to reduce housing subsidy in total and certainly to reduce the proportion of it flowing through local authorities.

Their strategy concerning rents, insofar as they have one, is to move decisively away from variant 2, to induce rents to float upwards towards variant 4 (partly to drive more people towards buying, partly to attract investment to the private rented sector and possibly with some version of variant 5 in mind) and to provide means-tested user subsidies as a form of market support so that variant 3 rents *do* approximate to variant 4. It is not only in the rented sector that housing-access problems have sharply increased. The 'Right to Buy' policy, and other initiatives to increase owner-occupancy have closed down other options and pushed hundreds of thousands into purchasing when their economic circumstances were not sufficiently robust to cope with servicing a heavy debt. The recession and mass unemployment of the late 1980s and early 1990s has exacerbated the effects. It seems fair to observe that most of this looks like a reactive and ideologically bound set of responses to problems that are still clearly worsening. It does not look like a reasoned long-term housing-finance strategy.

SOME ECONOMIC AND SOCIAL CONSEQUENCES

There have been a number of immediate effects on public budgets. The cost of Income Support payments to meet mortgage interest increased in 1990 prices from £64 million in 1979 (98,000 claimants) to £554 million in 1990 (310,000) – a real-terms increase of over 865 per cent (*Roof*, September/ October 1991, 12). Arrears on mortgage-loan repayments are likewise rising fast. The Institute for Employment Research at Warwick University has been monitoring repayment arrears since 1985. There have been dramatic recent increases. Arrears of twelve months or more stood at 21,580 in June 1990 and at 103,880 in March 1992. Those of six to twelve months rose from 87,790 to 186,200 over the same period (Ford, 1992). So over 290,000 house purchasers, about one in every thirty with a mortgage, are experiencing the fear of repossession.

The systematic closing-down of the renting option and the sharp move from 'social' to 'market' criteria at stage 4 (the allocation stage) of Figure 3.1 have produced clear adverse effects for those least able to compete in the housing market. Table 9.3 shows the changes over the 1980s in one of the key indicators of housing shortage and stress – homelessness.

Table 9.3 Homelessness in England and Wales 1978/9–1990/1

	1978/9	1983/4	1988/9	1990/1
Applications as homeless	113,433	137,677	177,149	203,102
Accepted as homeless	53,227	N/A	N/A	95,864
Put in Bed and Breakfast	7157	10,726	16,535	21,109
Put in hostel	8176	9835	12,606	14,623
Rehoused in LA dwelling	27,288	N/A	N/A	22,238
No. of household/days in Bed and Breakfast	318,702	717,535	1,211,045	1,708,596

Source: CIPFA, *Homelessness Statistics*

The number of households accepted as statutorily homeless has increased by over 80 per cent over the decade, and the use of emergency accommodation of various kinds has increased much more than this. The key figure, from both a social and an economic point of view, is the 536 per cent increase in the number of household/days spent in bed and breakfast accommodation. This 'solution' puts immense strains on the families involved and cost over

174

£66 million in England and Wales in 1991–2. In fact the total cost of dealing with homelessness was £193 million (both figures are from the same source as Table 9.3). As many people have pointed out, if a capital sum that cost this amount per year to service could be borrowed and invested in new council house building it would go a long way to removing the homelessness problem altogether. But neo-liberal politicians don't see things this way.

LONDON'S DOCKLANDS – THE NEW RIGHT IN PRACTICE

There is no point in spending too much time reviewing the twenty-year history of the London Docklands redevelopment. This has already been done (see, for example, the two excellent early accounts by Hardy, 1983a and 1983b; Ambrose, 1986; Church, 1988; Brownill, 1990; Docklands Consultative Committee, 1990 and 1992; Coupland, 1992; and Ogden, 1992). The aim here will be to draw on the more anecdotal accounts of people who have lived close to the redevelopment process and who are well placed to assess its economic and human costs.

The London dock system stretched for seven miles or so downstream from London Bridge. The local economy of the adjacent areas on both sides of the river was heavily dependent not only on the handling of cargoes and storage of commodities for trans-shipment but also on the processing of a number of key imported materials, particularly a wide variety of food and drink. For one hundred years or more the London dock system formed the world's busiest port, the mercantile heart of the British Empire and, in someone's phrase, 'the Emporium of the World' (Ellmers and Werner, 1991). Well into this century, in the inter-war and post-war decades, the dock system continued to flourish, but by the 1960s far-reaching changes were already on the horizon. The British Empire, that anachronistic collection of captive raw-materials producers and markets, began to fall apart and London began to lose its share of world trade. An over-capacity in upstream dock space and warehousing became apparent. The long-haul passenger trade quite suddenly deserted the ocean liners for the jet aircraft. Finally, the growth of containerisation as a form of shipment placed a premium on more accessible deepwater locations downstream. The tidal upstream docks were all closed between 1967 and 1981, and the heavily docks-dependent local economy collapsed. The largest redevelopment opportunity in Europe presented itself within easy reach, at the western end, of the previously hemmed-in City of London. But how was the redevelopment to be planned, financed and implemented, given that five relatively impoverished London boroughs shared the affected areas and that massive infrastructural investment would be required to get the area ready for redevelopment?

By the late 1960s the planners of the recently formed Greater London Council were working on the problem since the area had great significance for the overall strategic planning of London. Development undertaken in this 'windfall' area could well reduce the pressures in the 'Green Belt'. The 1969 Metropolitan Structure Map zoned almost all the area for industrial and commercial use but this was perhaps over-optimistic about the amount of investment that could be attracted to the area, and it took insufficient account of the need for more housing and recreational facilities. The various landowners had different priorities. The Port of London Authority wanted to maximise the proceeds from its considerable landholdings by selling sites for speculative owner-occupied housing and other high-value uses. Tower Hamlets Borough, adjacent to the City, saw advantage in increasing its rates base by attracting some high-value uses. Newham Borough, further downstream and with fewer chances to attract commercial development, wanted to increase its housing stock and produce a better social integration between the various parts of the borough separated by the Royal Docks system. Southwark Borough, one of the poorest in London and with growing pressure from articulate community groups, wished to gain revenues to update its housing stock and provide better service delivery in a number of spheres. Apart from the problem of competing agendas, an immense amount of infrastructural investment, at least £1 billion in early 1970s prices, would be required to drain the docks, clean up the heavily polluted land and improve the utilities and transport facilities of the area.

The competing aims and strategies for the area were articulated in a stream of consultants' reports and other less formal documents during the 1970s. Travers Morgan produced a report in 1973 offering a range of options; several of the Home Office sponsored Community Development Projects offered competing scenarios; and the Docklands Joint Committee, with representatives from the GLC and the five boroughs most concerned, produced the 1976 London Docklands Strategic Plan. This envisaged a twenty-year development period using a high proportion of public-funding input. But at this precise time public expenditure, especially by local authorities, was under heavy attack as a consequence of the terms of the International Monetary Fund 'rescue' of sterling in 1976. This reduced the possibility of any large-scale public-sector-led attack on the problem. In all, for a variety of good reasons, redevelopment was proceeding only slowly by the end of the 1970s.

The Conservative government elected in 1979 came to power pledged to deal with the redevelopment problems here and elsewhere by means of Urban Development Corporations. These were to be centrally appointed and funded bodies with the brief of cutting through the bureaucracy of local authority planning and regulation and bringing about a 'market-led' regeneration. The policy also brought about a sharp shift in the NDA:DA balance of power at most stages of Figure 3.1. Under the Local Government, Planning and Land Act, 1980 the corporations were given special land-acquisition and development-control powers which meant they, rather than

176

the previous statutory plans of the boroughs, controlled the pattern of development. In fact 'controlled' is the wrong word because the strategy underlying most of the London Docklands Development Corporation (LDDC) policies was to acquire sites, infrastructure them and then let them go to the highest bidder so long as the use proposed would not have adverse effects on adjacent site values. Visually the results are bizarre, but by its judicious land policy the LDDC was successful, up to the late 1980s, in raising the general level of development-land values from zero or even negative up to a highest price recorded of £18 million per acre for commercial development.

The strategy incorporated in the 1980 Act and implemented by the LDDC embodies many of the neo-liberal beliefs discussed in the previous chapter. The role of the state is confined to 'enabling' and supporting the market with 'kick-start' subsidies, for example the ten-year fiscal concessions in the Enterprise Zone in the Isle of Dogs. One of the main criteria set up to judge the LDDC's success was the 'leverage ratio', the amount of private investment attracted per £1 of public money spent. Private investors were attracted by a combination of good marketing, firm forward plans concerning infrastructuring, attractive land prices and a 'flexible and responsive' approach to development proposals. In fact it is fair to say that the size of the bid for a site, rather than conventional planning considerations, has become the main determinant of development outcomes. Virtually all development proposals have been underpinned by capital-accumulative motives rather than social criteria. The agency facilitating the development, although notionally 'public', is firmly in the NDA part of Figure 3.1, since the board members are appointed by the Secretary of State and individual development decisions, even very large ones, are not subject to participation or approval by local residents. The only line of democratic accountability is via the Secretary of State, who is ultimately subject to national electoral mandate

THE FAT CANARY – A DENIAL OF REALITY

The results of thirteen years of this strategy need not be reviewed in detail here since they have been well discussed in the literature noted above. The overall effect has been an extreme case of the speculative supply-side 'lumpiness' discussed in Chapter 4. A massive speculative-development boom occurred in the mid- and late 1980s, primarily of offices to let and housing for sale. Commercial/industrial floorspace completed or under construction in the Docklands as at March 1992 amounted to 28.2 million square feet (LDDC *Key Facts and Figures*, March 1992). If one includes 'planned' and 'potential' development, the effect is to add an additional City of London to the stock available for letting. At the same date 17,082 housing units had been built or were under construction. A very sizeable proportion of this development remains unlet and unsold – partly because the recession

177

Plate 9.1 Some of the 17,000 housing units built in London's Docklands between 1981 and 1992 under the revelopment strategy of the London Docklands Development Corporation. This new-build development on the south shore of the Thames, completed in 1990, has encountered very difficult marketing conditions. The two- and three- bedroom flats and houses are available at between £80,000 and £125,000 (1993). The builders, Fairclough Homes, offer a number of sales incentives including the 'Activator' scheme where the purchaser rents the property for four years at a low fixed rent then buys at a pre-agreed price (see Chapter 5 for a fuller account). Photograph by Keith Hunt.

which began in the late 1980s reduced the demand for business premises, and partly because the Corporation of London, as we saw in Chapter 4, fought back and generated a development boom of its own. Consequently a number of development companies, including both commercial developers and major housebuilders like Kentish Homes, have over-extended themselves and gone into administration.

Olympia and York, the world's largest property developers, built the Canary Wharf complex on what used to be the West India Docks. The scheme includes over 12 million square feet of office and retail space, much of it the 800-foot tower that now dominates London's skyline. The development costs have been variously estimated as between £4 billion and £8 billion. Since 79 per cent of the development is within the Isle of Dogs Enterprise Zone, this proportion of the capital cost attracted tax allowances which are estimated to have cost the Exchequer over £1 billion. As is well known, a large proportion of the new commercial space in the complex remains unlet and the developers are in trouble in the biggest possible way, with debts around the world running into several billions of pounds. This chapter will conclude not with yet more

Plate 9.2 Looking across the River Thames with the development shown as Plate 9.1 on the right. The Free Trade Wharf development of 169 luxury flats was completed in 1990, although some were sold 'off plan' in advance. All have now been sold. Prices ranged from £107,000 for a one-bedroom flat up to £190,000 for larger flats. All flats have a balcony, most with a good river view. Facilities include a leisure centre and pool, a sauna and a gym. There is an underground parking space for each flat, a 24-hour porterage service and careful security arrangements. The annual service charge for these facilities ranges from £1300 upwards, depending on size of flat. Photograph by Keith Hunt.

factual information about a story that will run and run in the media, but with the observations of three people who have stood close to the redevelopment process – a local councillor, a general practitioner, and one of Britain's most perceptive property consultants.

In a trenchant foreword to a recent assessment of Canary Wharf (Docklands Consultative Committee, 1992) Councillor Dave Lawrence of the Isle of Dogs Neighbourhood Committee sums it up as follows:

> From every viewpoint, Canary Wharf is a 1980's concept which arrived in the 1990's, long after its time had gone. It could hardly have arrived at a worse time, coming amid a frenzy of office development in London that bore little relationship to demand. In central London as a whole there is about 40 million sq. ft. of empty office space. This oversupply ... was exacerbated by a recession which was particularly severe for financial institutions.
>
> Canary Wharf remains 40 per cent unlet, costing Olympia and York millions in debt repayments every month.... O and Y were able to

179

persuade Margaret Thatcher to commit a massive package of public investment to improving Docklands transport links at the expense of other much needed measures elsewhere in London. These commitments, together with the availability of cheap land and the Enterprise Zone tax breaks, has led to Canary Wharf being one of the most heavily subsidised urban regeneration projects in Europe.

Even now, planning permissions are being granted to build yet more speculative office development in the Isle of Dogs. This surely is a denial of reality, and the overriding lesson we have to learn from the Canary Wharf fiasco is that we cannot again let the ambitions of developers divert scarce public resources on such a scale, while leaving the social conditions of local residents in surrounding communities largely unchanged.

Canary Wharf and the many other large-scale developments may well be anachronisms, because if and when there is again enough demand for office space to take up over-supply on this scale, perhaps in the next millennium, a different type and specification of building may be required. It may be worse than that. The 'nightmare scenario', given the rapid advances in information technology, is that a lot of the activities these buildings were designed to accommodate could theoretically be carried out in the back seat of a Rolls Royce. The amount of public subsidy that has gone into supporting these megalomaniac projects completely exposes the neo-liberal rhetoric about a 'market-led' regeneration free from state 'interference'. If state 'interference' had not occurred in the form of land at below market price, an undertaking to finance both rail and roads improvements in the Master Building Agreement and the massive tax concessions noted above, Olympia and York would probably not have gone ahead with Canary Wharf. Furthermore if additional 'interference' in the form of an expensive extension of the Jubilee underground line does not materialise soon, there appears to be no hope at all for the development. It is precisely the *absence* of any proper planning framework relating the evolving pattern of land uses to transport and other infrastructure that has compounded the development's problems. So far, despite all the misguided public financial support, Canary Wharf has failed in its own terms and has bankrupted its promoters. Meanwhile many people in the surrounding areas live in scandalous conditions, as the data in Chapter 2 indicate.

It is in these conditions that David Widgery practised medicine for twenty years until his untimely death in 1992. General practitioners are uniquely well placed to assess the effect of the built environment on people and when a doctor manages, amid all the other pressures, to write about these issues (Widgery, 1991) one should listen with care. When Dr Widgery first went to Limehouse to practise, activity in the docks was not far past its 1964 peak, when over 61 million tons of cargo were handled. In the early 1960s 50,000 people were employed on dock-related work and a further 200,000 in dependent industries. There was a strict labour hierarchy – 'The workers ought to know their place. But they had a place.' But the writing was very much on the wall. By the time he died the vast majority of the 250,000 jobs had gone and unemployment and worsening living conditions had taken their toll. Most had lost their place, in several senses. In the part of Tower Hamlets where Dr Widgery practised, unemployment is 30 per cent or above and rising. Morbidity and consultation rates are among the highest in Britain. There are unusually high proportions of ethnic minority populations – Chinese, Bangladeshi, Vietnamese and Somali. There is a high proportion of single parents, and a 30 per cent illegitimacy rate, and 25 per cent of homes are overcrowded. As the Tower Hamlets district psychologist told the borough's Health Authority Inquiry, 'The degree of isolation, depression and child behavioural problems are really the worst I have seen anywhere' (ibid., 15).

Widgery was clear about the processes at work here. In a passage which makes one wish he had found the time to produce textbooks on urban development he wrote:

> It is possible to describe this process in many ways. As the product of political economy and the way the City and deregulated finance capital have triumphed over manufacture to such a degree in late twentieth-century Britain. Or as a geopolitical imperative, impossible to halt and ludicrous to oppose, by which water transport and the industries and skills required to sustain it have been vanquished by road and air transport. It could even be seen (if you were to close your eyes almost completely and believe in the façade, and there are nowadays plenty of thirty-year-olds who seem to do this successfully) as part of a great leap forward to freedom (redefined as the market) against all the unwanted nonsense of the Welfare State and comprehensive education and council housing and municipal socialism.
>
> (ibid., 15–16)

He confesses to 'the grinding down of the optimism with which I came as a doctor to the East End nearly twenty years ago, into a kind of grudging weariness punctuated with bouts of petty fury' and concludes: 'I'm watching something die and I wish I wasn't. Perhaps the best that can be done is to record the process' (ibid., 16).

Widgery's book alternates between detailed accounts of his daily encounters with a wide range of medical conditions, many of them directly related to the overcrowded and squalid living conditions, and a solid documentation of those conditions (ibid., 118–21). By the late 1980s there were an estimated 6000 people in emergency accommodation in Tower Hamlets, many in hostels or low-standard short-life housing. As he rightly observes, housing problems are not just about people sleeping rough. They are also about access and affordability, about the cheapest available accommodation being above the combined means of two medium- to lower-paid workers and about the profligate waste of scarce public resources on 'solutions' that do not solve problems (see also Ambrose, 1992, 163–7). He points out that 'this ludicrous and cruel situation' is not the result of some natural calamity; it is the deliberate result of central government policies in the 1980s. Between 1981 and 1989 the number of new lettings available in the London boroughs fell by 41 per cent, new building fell by 91 per cent, and 17 per cent of London's public housing stock was sold off. Only the amount spent on bed and breakfast accommodation had increased (from £4 million in 1981 to £99 million in 1987): 'What is genuinely shocking is that, as an act of policy, over £223 million has been handed by the ratepayers to the hotel landlords without the gain of a single unit of permanent housing for the homeless and at the cost of such cruel damage to the lives of so many adults and children' (ibid., 119). He goes on to give a powerful account of trying to practise medicine in these conditions. For him the 1980s was the decade when 'one became hardened to the sight of hunched and dishevelled human bundles in doorways, disused buildings, even skips. Some enterprise, some culture' (ibid., 120).

And what human processes have produced these disasters? They are symbolised in the great central block of the Canary Wharf complex – the product of much busy-ness, much expensive research, much confidence and the commitment of immense sums of money. But

> for all the investment and activity even Canary Wharf remains curiously alien, an attempt to parachute into the heart of the once industrial East End an identikit North American financial district. To outflank the City of London to its east was implausible enough at the height of the deregulation boom of the mid-1980s. To do so in an era of major recession ... seems impossible.... But yet it hovers there, a gigantic Unidentified Fiscal Object, come to land on the old West India Dock.
>
> (ibid., 41)

What of the values and morality underlying these events? Widgery compares them with the action of one of his off-duty receptionists in going to a

Plate 9.3 (opposite) An icon for the 1980s. César Pelli's 800 ft stainless steel clad tower, part of the Canary Wharf complex. Begun personally by Mrs Thatcher in May 1988, and including over 500,000 square feet of space, the tower was ready for occupancy in 1991. Three years later the scheme was still less than 50 per cent let. The expectation that the development would 'create' 57,000 permanent jobs was a misreading of employment trends in the financial services sector serious enough to have driven the developers, Olympia and York, into financial crisis. For the development costs of about £1.5 billion one could build perhaps ten hospitals. Photograph by Keith Hunt.

This Property is
Protected by
Securiguard
Tel: 071 -
Working for London Docklands

patient's home to give some simple treatment:

> Proletarian decency over monetarist efficiency; one driven by compassion and the solidarities of work and neighbourhood, the other by the simpler calculation of profit and loss. There is no physical monument to what generations of decent working-class East Enders have created and given and made and suffered. But César Pelli the architect of the fat Canary tells us that 'A skyscraper recognises that by virtue of its height it has acquired civic responsibilities. We expect it to have formal characteristics appropriate for this unique and socially charged role.' Now that would be interesting to see.
>
> (ibid., 164)

Perhaps the enormity of the scale on which people have been misused in this area, and many others, in the 1980s and early 1990s, takes the issue beyond the orbit of the style of work and argument used by the authorities listed earlier in this chapter – an amalgam of carefully marshalled data and cautious critique. Perhaps the raw anger of a David Widgery is now the best weapon against the arrogance and myopia still being displayed by government across a whole range of social and urban policies.

ANOTHER LANGUAGE AND ANOTHER WAY

This anger can perhaps be allied with a further dimension of analysis and imagination. David Cadman has a deep technical knowledge of property-development processes and the dynamics of urban property markets (see Cadman and Austin-Crowe, 1991, etc.). For many years now he has built up a practice which has monitored trends in property cycles and has developed sophisticated forecasting techniques. His advice is consequently highly valued by some of the largest institutional investors in Britain. As an experienced property practitioner Cadman's reflections on the events of the 1980s, and on Docklands and Canary Wharf in particular, are well worth consideration. In an unpublished paper ('Visions of the City', 1993) he considers the immense amount of time, money and skill that has gone into the task of building and rebuilding a city like London and then wonders, looking across the London skyline, whether Canary Wharf is the best we can hope for.

Cadman argues that such gross errors arise because we are using an impoverished 'language' – a language that concedes an 'exclusive reality' to the terms used in finance and accounting and thus narrows down the limits of the debate. He argues:

> The mistake that we made was twofold. Firstly we saw the city as no more than a market place and secondly we saw only the physical city. In both cases we failed to understand, and continue to fail to

understand, that the real life of a city comes from a realm beyond mere commerce and that it is brought to life by the lives of the people that live and work within it, those 'patterns of events' that Christopher Alexander refers to in his book *The Timeless Way of Building* [from which he then goes on to quote].

... the very 'language' used in the development of Canary Wharf is inadequate because it is the language of abstractions – of calculations of price and cost set out on paper, detached from the land. Although it purports to describe the 'real world', this language is in fact a language of fantasy, a language of make believe or in this case perhaps of nightmare. I spend a great deal of my time dealing with these abstractions and I have seen how powerful they are and what mischief they can do!

(ibid., 2, 14–15)

So is there another way for Docklands, another 'story' told in another 'language'?

Cadman believes that there is and that this other way would be rooted in respect for the land. This in turn requires a respect for the myths and belief systems of the many past and present cultures that recognise the needs of the land. He quotes from the belief system of a tribe of North American Indians that one should step carefully upon the land because the faces of future generations are looking up from the earth waiting their turn for life. Perpetually to take from the land produces a progressive degradation. For it to remain 'in good heart' it must be rested and refreshed (Seymour and Girardet, 1990). Might what is clearly true of agricultural land also be true of developed land? The Dockland areas of London have been heavily worked for one hundred, two hundred or three hundred or more years, depending where one looks. They have been artificially transformed, heavily polluted, loaded with goods and used for a wide variety of intensive wealth-accumulative activities. When this cycle of activity was finished, as it was by about 1980, how might we have 'sanctified the land' and enabled it to become refreshed?

The solution Cadman proposes is startling: 'let us suppose, then, that after clearing away the debris of the past, a patch of ground had been carefully chosen, and there erected a great tent-like structure, resting lightly upon the ground.... This place would have been a place of celebration ... a place for festivities and entertainments' (ibid., 15–16). It would have been built largely by local people and over time would have spawned small shops and cafés, a local school, a health centre, offices, workshops and shops. It would generate a localised interrelated pattern of activities encouraging people to walk or cycle from place to place. The 'pattern of events' that places are really for would take shape and give life. Cadman points out that such a process of development, which we might perhaps call 'organic', would not have cost the

185

£10 billion reported to have been spent so far in Docklands. The houses and offices built would not be empty. 'And would the land really have been worth less?'

It would be profoundly wrong to dismiss this line of argument, grounded in human activity and need, as utopian. If anything is utopian it is Canary Wharf. For most of the history of human settlement the process of development has been something like that described by Cadman – a process of the gradual accretion of buildings and uses as need has arisen. The large redevelopment schemes that have punctuated this history – formal Baroque squares and avenues, London's Regent Street, Haussmann's Paris (see Mumford, 1966) – have not been on a scale to disconnect them structurally from the city around them. Even the megalomaniac excesses of Versailles, Hitler's city schemes and Ceausescu's palace in Bucharest at least had visual consistency, intended users and a political purpose, however disreputable. It has taken the Docklands redevelopment to combine a rate of construction massively ahead of demand, consequent financial disasters, immense public subsidy, the displacement of thousands of people, inaccessibility and visual confusion on this enormous scale.

Perhaps the outcome is not surprising. The LDDC's brief from government was apparently something like 'build as much as possible and the bigger the better' (taken from an interview given in 1989 by a member of the LDDC's strategic staff and quoted in Garvey, 1991). It was a very silly brief and betrayed much ignorance of urban process. Human activities generate needs for buildings, not the other way round. Of course speculative building has always occurred but not on this scale, not when the developments are so isolated from the rest of the city, and not when it looks as if many of the activities they were built for may be carried out elsewhere in the future. Only ideologues incapable of grasping that one of the roles of state regulation and planning is to protect the property market from its long-understood tendency towards speculative self-destruction could have presided over an error of this magnitude. Paradoxically the Docklands episode could in the end carry the benefit of providing a lesson, if an expensive one, in how *not* to manage urban redevelopment. One might even be pleased were it not that so many people have been hurt along the way.

Part IV

LESSONS
The widening of perspectives

In this final part of the book we look both outward and inward – outward to practices elsewhere that seem to offer greater possibilities for users' involvement in the control of the local built environment, and inward to see whether and how such practices might address the problems of loss of democratic accountability and growing inequality identified in Parts I and III of the book.

In Chapter 10 a number of international comparisons are made, including a consideration of democratic and participative practices in the Swedish and Danish housing systems. There is also a brief account of the activities of a British advisory team seeking to advise on the reform of housing production and management practice in Bulgaria – a country where people are seeking not so much to regain as to gain a degree of participation after living under a system which for over forty years has exerted highly centralised and bureaucratic control over the built environment.

In Chapter 11 we consider a number of issues raised by the earlier chapters – for example whether greater user input into decision-making processes would help to produce a more humane and equitable built environment given the loss of formal electorally based accountability noted in Part III. If it would, might the strategies and practices discussed in Chapter 10, and the attempts to secure greater user control over local environments which underpin them, represent a way forward in Britain?

10

HOW IT WORKS
ELSEWHERE

The London Docklands redevelopment provides an extreme case of the removal of local users' and residents' control over the fashioning of the built environment. Three underlying themes seem to characterise what has happened there – an externalised setting-up of the development 'ground-rules' allied with large-scale financial support direct from central government; a speculative regime for the individual development projects; and a massive scale of operation. It has been argued that to a degree these characteristics are predictable manifestations of the neo-liberal political framework in which the events have taken place. The LDDC is a centralist body because the Thatcher administrations could not expect locally elected administrations to carry out the damaging transformations required, and because in any event there has been a generalised drive to undermine the power and financial capability of local authorities since the fiscal crises of the mid-1970s. Individual development decisions have been speculative rather than planned because, as the discussion in Chapter 8 implies, it is fundamental to neo-liberal thought that investment decisions *should* be taken by the risk-takers themselves, untrammelled by any external authority or regulatory 'planning' framework. The scale has been massive partly because of the sheer size of the 'hole' to be filled but also because multi-national capital, attracted by the intriguing combination of heavy state support and a 'free for all' development regime, thinks and works at a far larger scale than local authorities and local citizens' groups.

Material presented in earlier chapters has shown that this way of fashioning new built environment has so far been a resounding failure when judged by widely diverse criteria. Was it inevitable that this redevelopment, standing as a symbol for many other less dramatic schemes, should be carried out in this way? Clearly not. Over the past fifty years many large-scale schemes have been carried out differently in Britain, and much European development, especially of housing, is carried out using other approaches and practices. In this chapter we will consider examples of development carried out on fundamentally different bases, partly because consideration of alternative modes of property development and management gives an extra dimension of understanding to familiar ones, and partly because these might serve as a model for the reform of our procedures – which many would judge is long overdue, and which might help to avoid future disasters.

THE SWEDISH LAND-DEVELOPMENT ARRANGEMENTS

This section describes the Swedish land-development arrangements as they existed before the General Election of September 1991, which ended sixty years or so of almost continuous social democratic government. Since then the ruling Conservatives have begun to dismantle the arrangements (Elander and Stromberg, 1992). Time will tell whether this is a brief interruption in the long period of social democratic control or whether the pre-1991 set of regulatory and supportive arrangements governing the development process will be lost in the flurry of deregulation already evident in Sweden's drive towards joining the European Community. Either way there is much to be learned from the 'Swedish model' as it applied until very recently.

The key to the pre-1991 Swedish land-development system is the strong element of local regulation over land supply exerted in an overall framework of centrally managed support for development and supply-side building incentives. In terms of Figure 3.1, promotion (stage 1) is highly democratically responsive, and changes in the political complexion of a commune can quickly produce different development outcomes; financing (stage 2) is closely regulated by the Bank of Sweden; construction (stage 3) is carried out largely by private contractors; and allocation (stage 4) and management (stage 5) depend upon the form of tenure of the housing unit. So far as housing is concerned, speculative development was very rare up to 1991 and most development was carried out by contract for clients, whether these be municipalities, co-operatives or individual promoters for their own use.

The main regulatory/support element since the reforms of 1974 has been the State Housing Loan (SHL). This has made construction capital available at a subsidised low cost from banks, other financial institutions and the National Pension Fund via the bond market and, since 1985, the National Housing Finance Corporation. All three categories of promoter, on accepting an SHL, had to build to within specified cost limits. The development had normally to be on land made available by the commune, under the 'land condition' (the *markvillkoret*). This land might well have been in the commune's ownership for some time, or it might recently have been acquired either by negotiation or under local authority land-purchase powers at something approximating to existing use value. It was made available to those promoting development with an SHL at a fraction of the cost of land on the free market. Housing schemes promoted by municipalities and co-operatives

190

needed to be in accordance with the commune's five-year rolling programme of housing types required given the changing demographic and other characteristics of the local area.

The overall effects of the system are to provide effective supply-side construction incentives in the form of cheaper construction finance, to make development land available to promoters at a cost limited by the SHL rules, virtually to eradicate land speculation because the commune has a near monopoly over the supply of land, to remove many uncertainties from the calculations of the construction industry, and to give communes positive powers over the pattern of development by means of their control of the supply and price of key inputs. This differs sharply from the arrangements in Britain, where the planning system seeks to bring about desired outcomes by negative control over the supply of development land and by specifying the use to which land may be put, but where there is no built-in capability to regulate the supply and cost of other inputs or any means to prevent speculation in them before they are put to use.

How well have such interventionary methods worked in recent decades? One good test would be to make a comparative assessment of the performance of the British and Swedish systems in a particularly testing situation – matching housing supply to demand/need in a 'high-growth' situation. This has been done in a series of research projects carried out in the late 1980s at the University of Sussex in collaboration with colleagues in Örebro in Sweden (see Barlow, Ambrose and Duncan, 1988; Danermark and Vinterheimer, 1991; Barlow and Duncan, 1992; Barlow, 1993). The areas chosen for comparison were the county of Berkshire in England, nine communes in the E4 corridor between Stockholm and Uppsala, a group of municipalities to the south-west of Paris, and the metropolitan area of Toulouse. All four areas were marked by rapid employment growth in the 1980s, with particular emphasis on 'high-tech' manufacturing, financial and producer services, and research and development activity. To some extent these were replacing older industrial structures, so that the effect on the housing infrastructure in the areas was more than simply to call for an expansion of the stock already there. Highly qualified and paid labour was being drawn into the areas, creating a demand for 'executive' housing. For example employment grew by 139 per cent between 1980 and 1987 in the Employment Growth Zone (EGZ) of the Swedish study area, with an emphasis on high status, highly paid jobs. The rates of growth of both employment and population, the definition of the study areas and an account of the methodology used in the research projects can be found in the works cited (especially Barlow and Duncan, 1992).

The comparative 'efficiency' of the housing-provision systems in matching housing to employment growth was assessed using a number of criteria. In terms of sheer output per 1000 population locally, the Swedish E4 corridor produced the best result at 7.3 housing units, compared to Berkshire at 6.1, but this comparison is of limited value since it does not take account of the nature of the pre-existing stock. It is, however, worth noting that Berkshire was the only area where output slumped towards the end of the period – by 60 per cent between 1987 and 1990, indicating a higher incidence of output volatility. Earlier research work (Dickens et al., 1985; Duncan, 1986) has compared British and Swedish output-cost trends in the period 1965–80 and found that the Swedish housebuilding industry had achieved increased productivity and better cost savings than the British industry, which had been characterised by lower levels of capitalisation and innovation (see also Barlow and King, 1992). One general conclusion drawn from this earlier work was that since land speculation has been largely excluded by the regulatory Swedish arrangements, the housebuilding companies there have been obliged to compete largely in terms of the quality and cost of the product. By contrast, their British counterparts have, in some states of the market, been able to make more profit from land dealing than construction and this has reduced the incentive to concentrate on the quality and cost of the product. This difference may help to explain why housebuilding costs rose in real terms by 47 per cent in Britain between 1981 and 1987, compared to 24 per cent in Sweden (Barlow and Duncan, 1992, 136).

In terms of the second major element in the cost structure of housing production, the cost of building land, the Swedish arrangements produced a strikingly better result in the 1980s than those in Britain. Land costs for single-dwelling housing remained relatively low and stable in the E4 area, although there was some increase towards the end of the decade. In Berkshire, by contrast, housing-land prices rose by 436 per cent between 1980 and 1988. In terms of price levels, by 1988 the average land price for single dwellings was £301,000 per hectare in the E4 corridor and £1,222,000 in Berkshire. This is all the more striking because both the British and the Swedish study areas were metropolitan fringe areas where land-cost trends might be expected to be similar.

The difference appears to stem from the key role played by communes in Sweden as the purchasers and providers of housing land. In Britain of course the housebuilder seeks land on a highly speculative private market, where the vendor is able to achieve prices under the 'residual' land-pricing method which reflect rising house prices (see Chapter 5). These in turn derive from periodic, large-scale and volatile releases of house-purchase credit. So land-price volatility, which serves no useful purpose and has some negative consequences, stems from a lack of sufficient regulation not only in the land supply arrangements but also in the financial markets. One consequence of these differences is that whereas in Berkshire the land-cost element in house prices had risen to about 60 per cent in 1988, in the two French areas it was between 30 per cent and 40 per cent, and in the Swedish case perhaps 10 per cent (Barlow and Duncan, 1992, 139).

Another criterion of 'efficiency' in housing provision is the degree of consumer choice available in the housing market. Neo-liberal theory holds that deregulation in production and consumption will lead to greater degrees of consumer choice. Clearly at one extreme that is true. The highly regulated and centralised housing systems of the former Eastern Europe certainly delivered very little in the way of choice. But towards the 'mixed-economy' end of the spectrum, and in the case of the areas examined in the projects under consideration, the relationship between the degree of regulation and choice is rather different. In Berkshire in the 1980s about 85 per cent of new housing was built speculatively for sale, and this proportion had reached 94 per cent by 1989. There was very little production of 'social' housing to rent. But in the E4 corridor, in the same 'high-growth' conditions, there was a diversity of output; 38 per cent of the new housing was built by the communes for renting, 27 per cent was built by co-operatives, most of the rest was built for owner occupancy (half of it by self-promoters) and a small proportion was built for private renting. Clearly most choice in terms of housing tenure and type was being delivered by the most regulatory system (the E4 area) while the most market-led system (Berkshire) was offering very little consumer choice.

The Swedish arrangements carry the additional benefit, in social terms, that there is a less clear relationship between tenure form and occupational status than is the case in Britain. Thus there is less danger of the 'ghettoisation' of specific tenure forms. For example in the E4 corridor in 1990 over 27 per cent of 'professional/managerial' workers were in 'social rented' housing and nearly 22 per cent in co-ops. In Berkshire the corresponding figures were 1.7 per cent in 'social rented' and 7.9 per cent in private renting. This indicates that people of various employment-status levels can make choices in Sweden which are not available to anything like the same extent in Britain. Other findings of the research were that households in the E4 area spend lower proportions of income on housing, that Swedish built quality is generally higher than in Britain, that E4 rent levels had remained steady in real terms and that there had been nothing like the house-price volatility which had been the case in Berkshire.

Finally, in relation to one of the key themes of this book, there is a high degree of local autonomy in the Swedish case – although the signs are that this may be weakening (Elander and Montin, 1990). Different local political control led to different patterns of promotion of new housing (Barlow and Duncan, 1992, 166), although virtually all forms of housing were built in all nine communes. In Berkshire, changes of political control meant relatively little in terms of the pattern of output because conditions imposed by central government made it progressively more difficult for *anything* to be built except speculative housing for sale. It seems clear that local political preferences were being translated much more clearly into housing output in the Swedish case, while market-based agencies, albeit heavily assisted by the pattern of housing support, were determining outcomes in the British case. In other words, and in terms of

Figure 3.1, events at stage 1 and stage 4 were much more powerfully conditioned by DA agencies in Sweden than in Britain.

SWEDISH HOUSING ARRANGEMENTS IN PRACTICE – A CO-OPERATIVE IN OREBRO

The previous section has considered differing degrees of democratic control over the promotion, production and financing stages of housing and shown that local democracy means more in Sweden than it does in Britain. This section will give a brief account of user control (stage 5 in Figure 3.1) in the management of a co-operative block in a suburb of Örebro, a medium-sized town in central southern Sweden. The block is a 'collective house' (*kollektivhuset*) in the suburb of Ladugardsangen, about two kilometres from the town centre.

The block, which contains sixty-one flats in a four-storey format, was built in 1990/1 for the 1992 National Housing Fair held in Örebro that year. The first residents moved in in February 1992. Under the co-operative form of tenure households pay a 'price' which buys for them the right of occupancy of the flat or house for as long as they want it. There is no 'freehold' in the sense of a permanent interest in the land on which the block is built. If the household wish to move, they sell the occupancy right back to the co-operative at the purchase price updated by general inflation. In other words the housing unit meets the need for shelter, but fulfils no investment function since it holds out no chance of capital gain. The confusion between the shelter function of housing and its investment function, which has so bedevilled the British housing system for so long and brought about wealth shifts that are both concealed and regressive, has been avoided.

The initial 'price' in this case was set by Obo, the developing co-operative, at 1000 Swedish crowns (say £100) per square metre or about £6500 for a one-bedroom flat of sixty-five square metres. This price was set in relation to the development costs of the block. In fact the first group of residents decided that this was too high in relation to the market and they took the decision to reduce it to 1000 Swedish crowns *per room*, accepting that this reduction in capital payment to the developers would carry implications in terms of higher future rents. This represents a striking example of user control not only over questions of management but over central issues concerning allocation (stage 4 in Figure 3.1) and the mode of recovery of the development costs. Those who had paid the original price per square metre were reimbursed. The policy seems to have worked because at the present price people of quite modest income levels can meet the access cost of buying into the co-op, and as at one year after the opening only eleven flats remained empty.

The block contains flats of a range of sizes and rent levels. As an example, the rent for a one-bedroom flat of 65 square metres is 3900 Swedish crowns (£390) per month. In fact such a flat would actually be sixty square metres;

194

Plate 10.1 A view of the *kollektivhuset* in Örebro, Sweden which is discussed in the text. In the foreground is the children's play area, which includes imaginative facilities with a variety of surfaces. The area is well lit and is enclosed by the two arms of the building so as to maximise surveillance. Also visible are the communal greenhouse, the bicycle parking racks (car usage is discouraged as much as possible) and small plots for residents to cultivate or keep livestock. The building in the centre of the photograph is the communal restaurant. The view also includes housing that is owned privately, by the municipality and by a co-operative. No differences in quality are evident between these various housing forms. Photograph by Christer Pöhner.

the other five square metres exist in the form of 'collective space' for which one pays rent. This space includes a laundry room on every floor and a room with bigger machines for rugs and so on, a solarium, a sauna, a gymnasium, a table-tennis room, freezer space and a communal restaurant seating about forty people with adjacent kitchen and lounge which can be reserved and used as required. Outside there are small plots of land which can be let to residents for cultivation, a greenhouse, and areas where small livestock can be kept. The block has been designed on ecological principles, for example the hard surfaces are fashioned to collect as much rainwater as possible, and there are a number of rich-smelling 'compost rooms' with movable metal containers in which small worms are working hard to break down kitchen waste for use on the plots outside. In fact there is compost to spare and this is sold on the compost futures market.

All the flats are completely self-contained, but the 'collective' in the description means that one can, if one wishes, take part in a number of collective activities. The most important of these is the monthly meeting open to all residents. This is presided over by an elected chair, a woman in this case. The meeting has power to take decisions on issues such as rent levels, although these can be affected only marginally since the largest element in the rent paid to Obo serves to repay the capital costs of the block's construction. The rent level therefore reflects to a degree the subsidy arrangements under which the block was built. In addition the monthly meeting can take decisions about maintenance (especially whether to call in specialists or do it oneself), the allocation of the plots of land, whether chickens should be kept, cleaning rotas for the communal areas, and issues that arise from time to time such as car-sharing and baby-sitting arrangements. Decisions on these matters, and a wide range of social information and small ads, including some of the lonely hearts variety, are communicated by means of a monthly newsletter which is distributed to all residents.

Apart from enabling residents to exert a considerable degree of control over various aspects of the management of the block, the collective arrangements also fulfil some of the social needs. One resident is an academic at the local university who used to live in a 'villa' or single-family house. He got tired of the routines of owner occupancy, the debt encumbrance, the cost of maintenance, the mowing the lawn and the competition with other residents in terms of the 'display' of possessions. Life is much simpler in a flat which allows as much privacy as is wanted, where the maintenance is looked after collectively and where one can participate in a variety of activities as fully or as little as suits the mood.

TENANTS' DEMOCRACY IN DENMARK

Non-profit housing has formed a significant part of the Danish housing system for more than a century. Professional groups such as the Association of Danish Physicians promoted housing complexes for workers in the middle of the nineteenth century, while large employers also developed housing for their workers, especially in Copenhagen and a few other large towns. Towards the end of the nineteenth century the Workers' Housing Association and other similar voluntary bodies had produced good-quality social housing often on a common-ownership basis. Around the time of the Great War the principle of tenants' democracy in the running of both co-operatives and housing associations became more common, and in the decades following the Second World War there was a considerable expansion of the stock. Some of this housing, especially that built in the 1960s and 1970s was in large high-rise estates, which led to various social problems.

Most of the output in the 1980s has been two-storey terraced housing in estates of one hundred units or fewer, and often in well-landscaped surroundings. Today non-profit housing, with a high degree of tenant participation in management, accounts for nearly 20 per cent of the Danish stock, while about 60 per cent is owner occupied and the remainder rented from private owners or co-operatives. There are about 650 housing associations in Denmark, all largely independent of local authority control. They own and manage about 450,000 houses and flats. The associations are of three types: housing co-operatives, self-governing associations and joint stock companies. The differences, which are to do with their capital base and the composition of the managing committee and of the governing bodies, are not important for present purposes because associations of all types work towards similar social housing goals.

While non-profit housing in Denmark aims to provide for a broad spectrum of society, it also operates with three broad objectives – to provide good housing for vulnerable groups, at reasonably low rents and with a high degree of tenant participation in the management of the properties. Municipal authorities have the right to allocate about 25 per cent of housing in the non-profit sector, and this helps to meet the first objective. Each association is run by a committee of management. In the 227 co-operative associations the residents have the sole right to elect the members of the committee. Nearly all the other associations are of the 'self-governing' type. Until 1984 these were controlled and managed by outside bodies such as the local authority or a trade union. Some were controlled by housing management associations. These are non-profit organisations which offer assistance with development, administration and management to member housing associations. KAB is the largest of these organisations. Its share capital is now 75 per cent owned by the housing associations to which it provides services and 25 per cent by the KAB Foundation, which supports innovative housing-management projects concerning, for example, energy efficiency in housing construction and management.

The 1984 reform ensured that majority control of each association was effectively in the hands of the tenants, even though some proportion of the ownership remained with an outside organisation. That body, typically the municipality, can still nominate some members of the managing committee. Tenants on a housing-association estate can participate in the management by electing typically three to nine of their number to form a local 'section board' for a particular estate. Each estate must hold an open meeting for this purpose at least once a year, and each household, regardless of size, has two votes. The votes are secret and in writing. The agenda for each meeting must include the election of board members, the accounts for the past year, and the budget for the coming year. Other matters of common interest, such as maintenance, security, order and the provision of space for communal and recreational activities, may also be dealt with. Official guidance and support are available to section board members in the form of courses, seminars and written information from both KAB and the Danish Federation of Housing Associations. Other matters between tenants and the association, such as rent levels, rules of occupancy and conditions for vacating flats, are regulated by law and by a ministry-approved leasing contract. So long as the contract is properly observed the lessor has no right of eviction, but tenants have a right to leave on giving three months' notice.

Rent levels vary, since the largest element reflects the borrowing and building history of each association. Since 1982 new non-profit housing development has been financed by 'low-start' loans which are index-linked to 75 per cent of the inflation rate. This means that initial rents are lower than they would have been had the loans been of the conventional variety where the loan-servicing arrangements are 'flat' in monetary terms and the annual revenue impact falls away in real terms with inflation. These loans cover about 94 per cent of investment in construction, and the rest is supplied by loans from the Building Co-operatives Fund, which is funded by both central and local state, and from tenants' returnable deposits, which are set to cover 2 per cent of the development costs. Developments carried out before 1982 also receive some subsidy, which is aimed primarily at reducing the annual revenue impact of the capital financing costs. The Danish system also includes some targeted demand-side 'housing-allowance'-type subsidies which depend on income, size of household and type of flat.

If a decision having cost implications is taken at a tenants' meeting, the impact is felt specifically on the rents of the section taking the decision; there is no facility for 'pooling' the effects across the whole stock of the association (or across the whole stock in a municipality, as has been the case in Britain). This avoids the possible source of contention that one is paying for someone else's improvements and conveys a greater degree of autonomy to tenants locally because they live with the benefits and costs of the decisions they take. Section boards in turn participate in a 'common meeting' of the association. At this meeting common problems are discussed and the supervisory board or committee of management for the whole association is elected. Working to some extent in co-operation with the local authority, this committee oversees the general implementation of policy. Lettings and the day-to-day management of the properties (stages 4 and 5 in Figure 3.1) are in the hands of a manager and her/his department.

To what extent do these arrangements represent genuine tenant democracy? In 1989 the National Federation of Housing Associations initiated a nation-wide survey carried out by the Danish Institute for Social Research. Data were collected from all associations by means of questionnaires, and the material was supplemented by personal interviews with a representative sample of 1000 households in a major provincial town. It was found that 32 per cent of tenants had participated in a general meeting during the previous twelve months and that turnout rates were highest in small sections or estates. Of those not attending the most recent meeting, only 8 per cent did not know that the meeting was to be held, so clearly the arrangements are well publicised. About one-quarter of the local residents participate in leisure-time activities organised by the local section. Of the 1000 households interviewed, 62 per cent were satisfied with the work done by the section board members and 80 per cent were satisfied with the overall management of their association. Section board members, in turn, felt overwhelmingly that they had adequate influence on the committee of management and the overall running of their association.

Danish housing administrators are proud, it seems justifiably, of the experience they have built up in the field of tenant participation and in the more technical aspects of housing management – for example in energy conservation. Their practices are among the most innovative in Europe and they are at present providing advice, by invitation, in a number of East European countries which are undertaking housing policy reform. A British housing expert, Ken Walker, prepared a report in 1991 on tenants' democracy in Denmark. He concluded:

> Somehow, Danish tenants have made the leap from seeing their main task as being to prevent any increase in the rent, to one of responsibility for the social and economic climate of their area or section. This may, in part, be due to tradition, but it has also evolved from a positive effort on the part of Associations to devolve responsibility and control to the tenants.... Great emphasis is placed upon helping tenants and their representatives to extend their role to beyond that of protecting their interests as tenants.... I believe there is much to be learnt from them.... We learnt many years ago that we could not change the structure of an area simply by giving people bathrooms and inside lavatories. Tenant participation is not about rent level protection. It is about pride, self-help and economic revival.
>
> (Walker, 1991)

AN ASPECT OF HOUSING REFORM IN BULGARIA

This section makes no attempt to present a systematic account of the housing-reform process in Bulgaria. It is far too early to do this, since at the time of writing (mid-1993) most of the legislation that will provide the framework for changes in land-development practices is still at the discussion stage, the land-restitution issue has not been finally settled and national politics are still in a transitional rather than a settled stage. The section will focus instead on the work of the British Housing Advisory Programme which was funded by the Foreign and Commonwealth Office 'Know How' Fund for a two-year period (1992–4) to assist the housing-reform process by offering advice and technical support.

Bulgaria is the most southerly of the Balkan states and it formed part of the Ottoman Empire from the fourteenth century until its liberation by the Russians in 1977/8. It then became an independent principality and finally a kingdom in 1908. The subsequent history has been turbulent, with border disputes with Turkey, Greece and Serbia, a failed Communist coup in 1923, a succession of coalition governments and then the personal dictatorship of King Boris III from 1935 until the war years. In 1946 the country became a satellite of the USSR under prime minister Georgi Dimitrov, who managed the nationalisation of the economy and instigated a system of five-year plans. Todor Zhivkov became first secretary of the party in 1954 and prime minister in 1962, and was finally toppled in a largely bloodless coup in November 1989. Bulgaria was perhaps the most closely related satellite state in the Soviet system, with bonds based on linguistic and cultural similarities, gratitude for Russian help in the 1870s and, as a founder member of Comecon, very close trade links with the Soviet Union. In all, in its century or so of existence, Bulgaria has had little chance to establish fully autonomous political, social and economic institutions, legal frameworks and practices. It is currently emerging from forty years of centralist and bureaucratic socialist control and seeking to develop a 'market-orientated' society – with little in the way of 'social democratic' tradition and with the help of advice from a variety of national and international advisory and funding agencies.

Contrary to popular belief, many of the socialist states of East Europe had high rates of private home ownership. In Bulgaria in 1985 only about 18 per cent of housing was rented (Giorov and Koleva, 1990) and by 1991 more than 90 per cent of the housing stock was privately owned, most of it by the occupants who had taken out long-term loans from the State Savings Bank at 2 per cent interest. In Sofia and the larger towns much of this housing is in the form of flats, while in rural areas and small towns there is a higher proportion of single-family homes, some produced under the strong self-build tradition that was almost totally submerged by forty years of centralist control of key resources such as finance and materials.

Under socialism, and especially after 1958 when the construction industry was nationalised, the quality of housing produced left much to be desired. By 1976 only 14 per cent of new construction was privately carried out, mostly in the form of self-promoted housing using loans from the State Savings Bank and building materials purchased from the state construction companies (Koleva and Dandolova, 1992). Most housing was produced by the forty-one state-owned enterprises and approximately ninety municipal construction companies. Much of it was built as sterile high-rise estates, often with inadequate community services. But the record in terms of sheer volume of output was impressive. Production between 1977 and 1985 was averaging about 70,000 units per year – a very high figure for a country of 9 million population. And between 1975 and 1985 the average annual growth of dwelling-units was 2.18 per cent per year, compared to the household growth rate of 0.96 per cent per year (Hoffman, 1991). The result was a sharp improvement in space standards. Between 1965 and 1985 the average number of people per dwelling-unit fell from 4.2 to 3.3, and the average floor space per person in square metres rose from 11.1 to 17.8. The long-term tendency towards rural-to-urban migration as the country underwent rapid industrial-isation led to long waiting-lists in the major cities (over 92,000 in Sofia by the end of the 1980s) and a vacancy rate of over 20 per cent in rural properties.

Despite these insensitivities of output to need, housing conditions improved considerably for Bulgarians under the latter decades of socialism, and housing costs, including utilities, were estimated to be only 6.75 per cent of incomes, whether buying or renting (Hoffman, 1991). These low housing costs reflected the high level of central subsidy into housing production. A wide range of subsidies, for example on materials production, the provision of cheap land, and production and purchase finance at reduced interest rates are estimated to have cost altogether approximately 0.6 per cent of the state budget during the later 1980s (Hoffman et al., 1992). Furthermore, sig-nificant numbers of state-owned units were sold during the late 1980s at prices per square metre considerably lower than those in the private-transaction market. Since 1989 the International Monetary Fund has been seeking to ensure that no further direct subsidies to housing are included in the state budget. There has also been a deregulation of prices and rents, and housing costs have consequently multiplied, especially in Sofia and the larger cities. By 1993 a professional couple renting a flat in Sofia needed to commit one salary and part of the second salary simply to pay the rent. The allocation stage in the model shown as Figure 3.1 has moved sharply from 'social' to 'market' criteria. In terms of Figure 1.1, the sharply increasing inequalities in the labour market are bound to be reflected in inequalities in access to housing.

The main housing problems facing the country as it moves into a new 'market-based' era are to do not so much with an absolute shortage of stock, although some shortages exist in urban areas, as with the lack of variety and with issues of management and maintenance. Many of the blocks have superficial structural problems in the form of leaking roofs and windows. There is a growing problem of security and no adequate mechanisms or organisations for resident involvement in the management of the common spaces, both internal and external. As a result the areas between the blocks often look totally uncared-for. Most cities are served partly by municipal heating systems and these have been inadequately maintained both in the central plant and in the local equipment. In all, while the volume of output has been impressive there is a massive 'deferred maintenance' problem, partly because initial construction standards were sometimes poor and partly because there has been no proper monitoring of the stock and no investment in regular maintenance programmes.

Various other problems have to be addressed. The rented sector is extremely small and this, together with the high and to some extent speculative pricing of flats for sale, has made access to housing increasingly difficult for the 'socially weak' – a category swollen by rapidly increasing unemployment. A shortage of rented units may also inhibit labour mobility.

The reduction of central-state expenditure on housing support, coupled with the as yet inadequate revenue base of local municipalities, means that there is no evident source of investment in rehabilitation and new-build. There is also the drag on central finances of the large amount of money lent out by the state under the previous easy terms (2 per cent over long periods) at a time when the nominal value of money is being reduced by rates of inflation of 80 per cent per year or more. There is no established body of housing expertise in most municipalities (although the quality of the few experts that do exist is very high) and very little in the way of pre-socialist tradition and institutions to reinstate.

The British team that sought to address some of these problems in the two-year programme commencing in 1992 included twelve people, drawn from the private, public and voluntary sectors. It included an architect, two lawyers, a quantity surveyor, a local authority housing manager, a planner, a land economist, the director of a voluntary housing organisation, an expert in development finance, and three academics. Several of the team had considerable experience of working closely with residents, especially in refurbishment schemes. The strategy adopted was to work at the local level with three chosen municipalities as well as with the minister and senior staff in the ministry responsible in Sofia and with the National Centre for Regional Development and Housing Policy. The essence of the programme was to encourage the initiation of specific construction and/or rehabilitation projects, chosen by each municipality in the light of its needs, and to pass on expertise about Western institutions and practice as problems arose in the development of each project. The advisory work was therefore at a 'micro'-scale to complement the advisory programmes being offered by teams such as that from the US, which concentrated on macro-economic issues.

The municipalities chosen in the initial competition to take part in the programme were Elena (population 8000, in central Bulgaria), Popovo (population 20,000, also in central Bulgaria) and Vidin (population 80,000, situated on the Danube in the extreme north-west of the country). Popovo had suffered from a severe earthquake in 1986 and was already benefiting from a number of central-government programmes to deal with the aftermath. The advisory team worked closely with a project team in each municipality and with liaison staff from the ministry and the National Centre in Sofia. The projects selected by the towns, in collaboration with the British team, included a small self-build scheme designed primarily to house lower-income people (in Elena), the 'mixed' development of a prestige city-centre site by means of a private/public partnership (Popovo), and the development of a purely residential scheme with greater degrees of resident participation than was formerly the norm (Vidin).

As the collaborative work developed, the British team sought to convey two underlying principles – both germane to the central themes of this book. One was that if there is very little or no public money, then ingenuity and new practices have, to the extent that it is possible, to serve instead. The second is that if community and resident participation can be achieved, and if the people who have a direct interest in improving their homes and immediate environment can be drawn into the process, then the 'release' and organisation of individual and group energies can help to bring about desired outcomes even in periods of great financial stringency. Thus the problem was not conceived of as simply or even primarily technical – although a constant stream of advice was offered concerning matters such as the workings of the land market, construction contracts and procedures, design issues, feasibility and market studies, and so on. An important aim was to assist in the activation, or re-activation, of citizen activity following forty years of repressive government during which very little in the way of self-help had been possible and 'the state' had provided. Related to this was the need to foster 'voluntary-sector' activity and to help in devising and implementing the legal entities and the participative attitudes and practices that could make 'people involvement' a reality.

The main hope of the British team was not, therefore, that these small projects would transform the urban environment of Bulgaria, but that ideas about self-help and group initiative, which have proved potent elsewhere (see the Swedish and Danish material above), would take hold and be implemented. It was also the hope that events in the three towns would be watched and that the new practices, especially involving community and resident participation, would be incorporated into the development process in other towns and cities. As a means to achieving these aims there have been a number of exchanges of visits. Officials and politicians from the three towns have had direct experience of a variety of development processes in Britain, and the British team has learned a lot about Bulgaria and the problems of social and economic transition.

The Elena group has seen a number of self-build activities in Britain and has been very quick to learn how voluntary-sector agencies can assist the process both financially and in other ways, how the local political process can be geared to supporting the activity and how important it is to seek consensus between local politicians, experts and people. The Popovo group, led by a very able deputy mayor and former banker, have learned a lot about the finances of 'mixed development' in a largely capitalistic land market and the way in which the long-term interests of the town can be protected in a 'partnership' scheme. They are implementing this knowledge in their negotiations with the private-sector developers carrying out the work. They have also taken steps to publicise their intentions and to elicit citizen response to the various alternative schemes. The Vidin group, while continuing with a residential scheme that was devised before 1989, have changed the design from high-rise to low-rise and have taken into account the need for much greater resident participation in the design, allocation and management of the properties. They are also working to implement a management agreement drawn up in collaboration with members of the British team.

The development of the Elena project is perhaps especially interesting. A charitable, non-profit, apolitical foundation has been set up, with a number of prominent citizens as trustees, to oversee the development of twenty or so single family houses with gardens – a type of housing not built in Bulgaria for many years. The management of the construction and the sub-contracting will be in the hands of a company to be jointly owned by the municipality and private investors. The site has been given by the municipality, who will also provide all necessary infrastructure and a capital grant of 1.5 million lev (£37,500). It is estimated that the total construction cost can be kept to about £15,000 per unit with the help of savings on material costs because of the charitable nature of the enterprise and the input of free labour by the self-builders. A donation has been raised from a British source, and the Bulgarian minister has agreed to contribute a more than matching amount for donations of this kind. A limited number of the houses will be sold to provide more capital finance for the rest. If possible, these sales will be to poorer people eligible for some state compensation for the real-terms decrease in their bank deposits. The remainder of the units will be rented. The municipality will be granted some nomination rights in exchange for their input of public money and land, and the Foundation will nominate the rest of the tenants. The scheme combines social purpose, small scale, well-designed housing and mixed financing in ways that would have been quite impossible under socialist arrangements.

It is too early at the time of writing (halfway through the two-year programme) to make great claims about the success of the Advisory Programme. While the Bulgarian urban landscape might not yet be transformed, the British team is working on the assumption not so much that the pen is mightier than the sword as that the power of ideas, the friendly meeting of minds, the interchange of visits and the intensive workshop discussion will prove to be more cost-effective than the previous socialist strategy of producing an endless succession of panel blocks. Unfortunately improvement in the built environment on any significant scale will depend upon the development or otherwise of the general economy or upon attracting investment from foreign interests. Energies are now being put into attracting this investment, and some success has already been achieved.

11

WAYS AHEAD?

Chapter 2 of this book identified and quantified the considerable differences that exist between two of the poorest and one of the most prosperous local authority areas in England. It is clear that the disparities between richer and poorer areas, and people, are if anything widening. There is no evidence of any progressive redistribution of wealth and welfare since the late 1970s and much evidence pointing to regressive effects. For example in 1987 the total tax burden on the poorest one-fifth of the population was 36.9 per cent of income and on the richest one-fifth 36.6 per cent (Johnson, 1991, Appendix, Table 28) – a marginally regressive effect. One study (Brown, 1989) found that the poorest 20 per cent were 6 per cent worse off in 1988 than in 1979, and the number of people at the Supplementary Benefit level or below increased from 6,070,000 in 1979 to 9,380,000 in 1985, an increase of 55 per cent in six years. A report from an international agency (Unicef, 1993) showed that Britain is slipping down the world league in relation to children's health and welfare. This is related to the increase of 40 per cent during the 1980s in families in Britain who live below the poverty line (taken as 40 per cent of national median disposable income). Interestingly, those other 'advanced' countries who have espoused more *laissez-faire* economic policies have also done relatively badly in the world league. The US, for example, has 20 per cent of its children living below the poverty line.

Other work which has concentrated on policy changes under neo-liberal Thatcher administrations (for example Andrews and Jacobs, 1990) shows that the 'reform' of a number of state support systems, or more accurately their partial withdrawal, has had the overall effect of 'punishing the poor'. In two of the areas described in Chapter 2, Tower Hamlets and Knowsley, the effects of these national policies were evident in terms of housing stress, lack of employment opportunities and decline in the quality of the local environment. By the early 1990s the third area, Wokingham, was beginning to suffer the effects of unemployment and declining property values – as a result not of conscious policy but of the recession that overall economic management had produced.

In a sense it is unfair to present the growth in inequalities as a criticism of the neo-liberal adminstrations of the 1980s and early 1990s, because they were the intended outcome of the Conservatives' policies anyway. In a book he co-authored in 1978, Sir Keith Joseph explicitly rejected the Beveridge-based philosophies and policies of the post-war era (Joseph and Sumption, 1978). As Riddell put it, 'Joseph ... made the rejection of egalitarianism a central plank of his counter-revolution' (Riddell, 1991, 150). There was a clear rejection of any notion that 'fairness' should be a factor influencing the distribution of income or wealth. This could only interfere with the optimal working of markets in 'free economies'. Not only that, but there was a rejection also of the very definitions of poverty used above. Poverty should be measured in terms of some absolute level of need, not by reference to the incomes of those who are not poor. This seems a curious assertion. There appears to be little else in social science, or natural science come to that, which is measured in relation to some notional absolute rather than in terms of relativities.

THE DRIVE TO DEREGULATION AND SOME OF ITS EFFECTS

Underlying these increases in inequality, and structurally connected to them in complex ways, lies the all-pervasive trend towards the deregulation of economic activity. As the discussion in Chapter 8 made clear, the four Conservative administrations since 1979 have been guided by the argument that Britain can best maintain her competitive position in world markets by 'lifting the burden of the state'. This philosophy has been translated into policy by deregulating the activities of markets, reducing planning controls and shifting the provision of key services such as health, housing and transport from the public sector to the private or the 'quango' sector. As part of this drive the intention has been to transform the role of local government from that of provider to that of an 'enabler' or contracting agency. Examples of this trend were given in Chapters 6, 7 and 9.

As one advocate of this general philosophy put it (Howe et al., 1977), this transformation is required because the state 'consumes' the nation's wealth; central and local government spend taxpayers' money but 'produce nothing'. But this is nonsense. From the earliest days of levying taxes to wage war or repair the city wall, the revenues raised and spent by the state produced *something* even if only in the form of armaments. In today's complex mixed economies, state expenditure is a crucial element in providing not only defence, but also education, training, research, infrastructure, codes of law, policing and other property protection, fair-trading regulation and much else besides – nearly all of which facilitates the operation of markets and the orderly accumulation of capital by private entrepreneurs. Not only that, but state regulatory activity reduces the risk of speculative self-destruction, as its partial removal in the case of Docklands has convincingly shown (see Chapters 4 and 9).

It is true that there is some validity on both sides of most arguments. Anyone with a knowledge of town-hall procedures might well maintain that there is a constant need for vigilance to ensure that bureaucratic practice does not gum up the works – that, as Heseltine put it in the early 1980s when castigating planning, 'jobs are not locked up in planners' filing cabinets'. Similarly, the management of local authority housing was for many years far too paternalistic and maybe still is in some areas. These are reasonable criticisms and grounds for the constant re-evaluation of practice. But the neo-liberal ideology goes far beyond this and seems to call for deregulation everywhere, for market-led solutions to every problem and for a simplistic (and *Nineteen Eighty-Four*-ish) 'market equals good, state equals bad' approach.

The moral position has been inverted. In the 1870s Joseph Chamberlain was arguing that it was a civic duty to raise taxes and spend them on improving the condition of the Birmingham poor (see Ambrose, 1986, 104–6). Before him the preacher/politician George Dawson had maintained as part of his 'civic gospel' that the Christian duty to help others was best fulfilled by the secular mechanism of local government activity. Speaking at the opening of the Birmingham Free Reference Library in 1866 he stressed the 'moral and intellectual purposes' of such publicly funded amenities. By the 1980s those of a neo-liberal persuasion had redefined public spending as 'waste' – as a cardinal sin of government. The market, with its free interplay of buyers and sellers, its 'discipline' and its 'undistorted prices', was the only moral way for investment and production to be organised and the product to be allocated. Only in a 'free market' could consumers exercise their right to choose between commodities on the basis of 'undistorted' prices.

The drive towards deregulation during the 1980s, and the related move towards the privatisation of both funding-flows and productive activities, had a number of effects so far as the built environment is concerned. In Chapter 9 it was shown that the pattern of promotion changed sharply. Whereas in 1978 40 per cent of construction orders were placed by 'public' agencies, roughly those in the DA sector of the model shown as Figure 3.1, by 1990 this proportion had fallen to 27 per cent. Thus nearly three-quarters of all construction was by the latter date being initiated by organisations seeking to accumulate capital rather than serve need. The result of this change is reflected in the changed output pattern. In 1978 40 per cent of orders had been for housing; by 1990 this proportion had fallen to 20 per cent. Within the housing category the proportion of orders placed by public agencies fell from 34 per cent to 15 per cent over the same period. So there has been a double effect on the availability of 'social housing'. Less housing is being promoted and produced overall, and of that smaller quantity a much larger share is by profit-seeking organisations; in other words more is for sale or rent on the market and less for allocation on social criteria. Inevitably housing problems have increased for lower-income people. So what was being produced instead? The figures show that the output of privately promoted 'commercial' development, offices, shopping centres and so on, rose by 313 per cent between 1979 and 1990. While the connection between the neo-liberal politics and the built-environmental outcomes has not been conclusively demonstrated, it would be pedantic to seek to do so. It is perfectly obvious that there is a reasonably direct causal connection.

HAS DEREGULATION DELIVERED THE GOODS?

So we come to the next question. The deregulation, the privatisation, the change in the state:market ratio of activity and the changes in the output of new built environment that have logically ensued have all been implemented ostensibly in the interests of economic efficiency, national competitiveness and 'growth'. In Josephian terms they were to improve overall economic performance. The LDDC was to oversee the regeneration of Docklands by 'market-led' strategies so as to keep London upfront in the European financial league. Financial markets were deregulated and rent controls partially removed to stimulate housing activity. Areas like Berkshire were to be our 'high-tech Utopias' (Barlow and Savage, 1991), where growth was to be inhibited as little as possible by 'planning constraints' and other forms of regulation. Compensatory 'regional policies' for areas with less growth potential were to be swept away. Has it worked? Have the undeniably grievous effects on poorer regions and people been justified by gains in the output of the total economy? In the end will these benefit the poorer groups by the 'trickle-down' effects whose believed existence provides some moral justification for neo-liberal thinking?

The questions are counter-factual and cannot be answered. The post-1979 period cannot be re-run using different beliefs and policies to see whether things would have been better or worse. But one can point to indicative evidence of various kinds presented in earlier chapters of this book. The construction industry, for example, has since the late 1980s been experiencing one of its worst slumps since the Second World War. The value of orders placed in recent years has fallen sharply in real terms from the levels of 1987–9 and this has affected all categories of work except industrial property. As shown in Chapter 4, the boom in commercial development of the mid-1980s onwards led to a glutting of a number of property markets and to catastrophic falls in rents and reduced yields. By the early 1990s some of the biggest developers in the field were in deep trouble, and financial institutions, trustees of our pension and life-insurance rights, had lost billions of what in the end is our money. If this is simply a 'shake-out' of inefficient players in the market, it is a very drastic and damaging one.

Similarly in the housebuilding sector, as shown in Chapter 5, the deregulation of financial markets in the early and mid-1980s led to a rush to lend by the institutions, upward pressure on house prices, and over-extension by many thousands of borrowers scrambling to buy. The resultant calamities of widespread mortgage arrears, repossessions and a large unsold stock of houses have brought much private misery, enormously increased by both the state's Income Support bill and the costs of bad debts to lenders. The effect has also been to discourage building firms from undertaking new programmes. Housebuilders who heavily invested in land have experienced additional difficulties as land values first peaked and then slumped following the trends in house prices. It is no wonder that spokespeople in both the finance and the construction industries are blaming the government for continuing with the deregulatory policies which have led to these volatilities and for refusing to organise some counter-cyclical influences in the form of large-scale public programmes of housing or infrastructural investment.

Further signs that deregulatory policies have failed to deliver benefits in terms of better outcomes can be seen in the results of the three-nation comparative study of labour-market/housing-market interaction in 'high-growth' areas discussed in Chapter 10. Because of the complexity of the issues, and the dangers of reading too much into a single study, it is safer to view these results as suggestive rather than conclusive. But a very clear picture emerged that the pre-1991 Swedish development system, with the local municipalities playing a powerful role in land supply and the setting of building programmes, and the state supplying finance on advantageous terms, clearly out-performed the more 'market-led' and speculative British system. Comparing the development of Berkshire and Stockholm/Arlanda during the 1980s, the Swedish case showed a better rate of housing output, a lower incidence of house-price volatility, far less land-price inflation, a trading environment that induced constructors to concentrate on better building not better land speculating, a less close relationship between occupational status and housing tenure, and finally, and presumably hitting the New Right where it hurts most, far more choice for consumers.

Finally, and briefly because the story is by now so familiar, there is London Docklands. The slow and difficult search for a consensual solution that marked much of the 1970s was inhibited both by the variety of interests who all, quite reasonably, felt they had a stake in the process and by the sudden contraction in public spending following the IMF rescue package that occurred at the height of the negotiations. The chosen Thatcherite solution following 1979 was a state-appointed, state-funded quango equipped with draconian powers and a very unintelligent brief. The regeneration was to take the form of producing as many buildings as possible by attracting as much private-sector investment as possible at a very heavy public cost in terms of inducements in the Enterprise Zone and elsewhere. Decades of planning orthodoxy were ignored – the buildings were to come first, with little regard either for the level of demand for them or for the transport infrastructure that would be needed to service them.

How can this lapse from long-established land-development practice be explained? Partly by the all-pervasive distrust and dislike of local authorities, especially leftish inner-London ones, which is evidently part of neo-liberal ideology. But partly too perhaps because the 1970s approach had been basically a search for something like consensus. The quotation from Mrs Thatcher at the end of Chapter 8 explains much about the sharply contrasting strategy of the 1980s. For her the search for consensus implies 'abandoning all beliefs, principles, values and policies'. It is a way of 'avoiding the issues that have to be solved ... merely' to get agreement on the way ahead. This stands democratic politics on its head. The search for consensus is not about abandoning anything – unless it be totalitarianism. It is about *recognising* the variety of beliefs, principles, values and policies and then finding ways in which solutions can be hammered out that can be accepted by a reasonably wide range of interests and that therefore might work. Seeking agreement in this way is central to democratically based politics; it is not a 'merely'.

In terms of the overall results the Thatcherite market-led strategy, as applied to Docklands, looks so far to be a disaster in its own financial terms. As Chapter 4 made clear, the commercial property-development boom in the latter half of the 1980s produced an over-supply of offices in the longer-established areas of London. This, allied to the prolonged business recession that struck the south-east in the later 1980s, led to the sharp falls in office rents detailed in the same chapter. In fact office rents in central London fell catastrophically from an index of 645 in May 1990 to 369 in May 1992 (May 1977 = 100). Shop rents and industrial rents fell too over the period, although not so sharply. But this was the very period when a vast amount of new space was coming onstream in Docklands. As at 31 March 1992, 21.7 million square feet of commercial/industrial development had been completed in the area, and another 6.5 million square feet was under construction (LDDC *Key Facts and Figures to the 31st March 1992*).

Partly this massive output flowed from the scramble to begin developments in the Enterprise Zone before the capital-allowance tax concessions ended in July 1991. Even without the effects of the recession it seems unlikely that the demand for offices and industrial space could have grown at this rate in the London area. Naturally enough a number of developers have got into debt-repayment problems and some have gone into liquidation. Their problems stem not so much from miscalculation or 'bad planning' as from the premeditated *removal* of well-tried procedures that seek to assess demand and exert some regulation on the rate of development in the light of these assessments. Neither the flood of development in Docklands during the 1980s nor the pro-development response adopted by the powers that be in the Corporation of London was carried out in the context of any strategic master plan for the respective areas. In the absence of any planning and regulatory framework of this kind, the market players all tended to act speculatively. The glutting of the market, the collapse in rent levels, and the bankruptcies are the result.

It is of course conceivable that as the 1990s proceed there will be a strong business recovery and much or all of the vast amount of commercial space in Docklands will find tenants at rent levels which produce the expected returns on the capital invested in the buildings. Whether or not this happens will depend on a number of factors, including London's future position in the hierarchy of international financial centres. One recent assessment (Coakley, 1992) is cautious about this. Competitor centres such as Tokyo and New York, reflecting the power of the Japanese and US economic 'blocs', are gaining primacy in some activities. In terms of London's position in the EC bloc, it seems that some continental banking systems are more in tune with the future liberal and universal direction of EC-wide banking. Britain's generally lukewarm stance towards European integration is unlikely to help in maintaining London's present primacy in a number of financial markets. The assessment concludes: 'It is not inconceivable that by the middle of the 1990s London will be struggling to compete with Paris and Frankfurt for financial hegemony within Europe' (Coakley, 1992, 70). If this bleak scenario does develop, the burst of unregulated and speculative activity that produced massive over-development in Docklands and the City during the 1980s may look, in retrospect, to be a misuse of public and private investment on a scale that neither sector can afford. At present it is difficult to see how this particular deregulatory episode has aided the national economy or Britain's competitive position in the world.

DEMOCRATIC ACCOUNTABILITY AND THE BUILT ENVIRONMENT

At various points in this book attention has been drawn to the de-democratisation of the processes fashioning the built environment. These trends have been evident in the changes in the NDA:DA ratio of activity in Figure 3.1, in the partial removal of land-use planning and in the loss of local authority power in special redevelopment situations where Urban Development Corporations have been set up. The Introductory Note indicates how widespread have been these tendencies. The problem is that the loss of democratic accountability in these ways is less dramatic, and less visible to the electorate, than would be other means by which a similar diminution of democratic power might come about. New legislation that specified a general election only every seven years, or which debarred 25 per cent of the population from voting would erode our cherished democratic rights to much the same degree as has occurred by the removal of large proportions of resource decision-making from the democratic to the non-democratic sector of Figure 3.1. Changes in the law in these respects would no doubt provoke much public outcry and debate. What we are seeing in relation to the built environment is a similar effect by more covert means and with less public challenge.

It may be as well, at this point, to consider whether environment-producing processes *should* be democratically accountable at the local level or whether they are best left to some combination of central *Diktat* and local market process. 'Should' is always a difficult word in social-policy discussions, since the values on which normative judgments are made are frequently left unstated. Nevertheless the outline of a pro-interventionary argument will be attempted. Advocates of a significant degree of regulatory activity, democratically accountable at a *local* scale, can raise arguments on a number of grounds. They can argue for example that there should be accountability in the way that locally raised taxes are spent. They can point also to the essentially public nature of the built environment regardless of who actually owns specific buildings, to the income and wealth-distributional implications of alternative housing and construction policies, and to the high social and economic cost of inappropriate mixes of development. They can argue, finally, that the future shaping of the local built environment should proceed on the basis of some participative and responsive assessment of the future demand for development, as distinct from assessments made by individual promoters for self-seeking reasons or by consultants with variable motives.

The 'general financial accountability' argument is simply a modern version of the 'no taxation without representation' demand that played so large a part in the claim for independence by Britain's North American colonies – which should act as a warning. Vast sums of publicly raised money are expended every year on changing the built environment. The combined cost to the public purse of housing support, the Urban Development Corporations, the road, rail and underground building programmes (and the balance of investment between them is a huge issue) and investment in airports, docks, physical infrastructure, and so on, consumes many billions of pounds annually. User input into the decisions made in many of these programmes is often derisory – to some extent forgivably so when the investment programmes have so long a timescale and are conditioned by so many complex factors. But the obvious difficulties in eliciting user responses, or even in building in some user influence over the investment decisions, should not justify a systematic move towards growing NDA control over the whole process – which is what we are seeing. Every pound spent from the public purse was either raised in taxation, borrowed for the purpose of achieving some public benefit or received from the sale of an asset that had been public. The nature of the relationship between the citizen and the 'democratic' state, if not natural justice, seems to call for the maximum possible electoral influence over the expenditure of these vast sums – especially since it is the human habitat that is being shaped.

212

This environment is also, as many have argued, common property in the broadest sense. Public urban spaces such as Piccadilly Circus, the Champs Elysées, the Ringstrasse, and so on, stand as visual and psychological symbols of their respective cities regardless of the private interests that own the properties that line them. 'No building is an island, entire of itself', to adapt John Donne. It is well understood that major redevelopment proposals in such sensitive areas will be a matter on which a broad range of interests in the city will feel they have a right to be involved. Less dramatically, the development of an out-of-town shopping centre carries implications for the 'life' of nearby town centres; a large new office building generates movement flows that affect the comfort of those already using the city; and the development of a peripheral 'high-rise' estate, perhaps with inadequate service and transport provision, may lead indirectly to social tensions that would not have arisen had the same number of people been housed in a more integrated manner in the town. In other words any urban area is an interactive system, in some ways an eco-system. Any significant change to the built environment carries 'area' and 'neighbourhood' effects in ways that the end-products of other accumulation cycles do not. It is consequently reasonable to expect a greater degree of democratic input into these particular accumulative processes.

A further argument for democratic participation in the processes that shape the environment is that their effects are not distributionally or environmentally neutral. The spatial configuration made by one category of buildings (say, housing) in relation to another (say, those where people work) and to a third (say, those where people shop) carries with it a complex pattern of costs and benefits. Some people are bound to spend more time and money than others on the movements required between the three areas. Furthermore, separating out different activities into major 'blocs', which has been one general result of land-use planning orthodoxies since 1947, is likely to increase the aggregate demand for movement within any urban area of given size to more than it would have been had uses been more spatially integrated. This has enormous energy-consumption, and ultimately eco-logical, implications. The low-energy-consumption city, and the one with the least inequitable movement-cost impact, is likely to be one where different uses are spatially well integrated. This is becoming easier to achieve as a greater proportion of jobs becomes 'clean' in terms of environmental pollution. In fact most air pollution may well come now not from industrial activity itself but from people making half-hour car journeys across the city to get to work.

The argument about the present and future costs to public expenditure of sub-standard built environment and declining economies needs only to be noted here – it has already been made unwittingly by the recent history of neo-liberal politics. At the national scale, financing unemployment has consumed immense resources and confounded attempts to force public spending down in real terms. In the poorer areas such as those identified in Chapter 2 the failure to invest public resources in the regeneration, or at least the humane transformation, of local economies and environments has produced personal stress and ill-health on an unmeasured but clearly enormous scale (see Hyndman, 1990). Under-investment in the upkeep of existing housing, both public and owner-occupied, produces sub-standard human environments in which the incidence of a whole range of conditions from hypothermia to asthma are far higher than the national average. The evidence has been summarised in a number of works (for example Byrne *et al.*, 1986; and Ineichen, 1993), and further evidence is available in work discussed in Chapter 9 (Widgery, 1991).

All these conditions require remedial treatment, and this all costs money. A more sensible approach might be to invest more money in redressing or preventing the environmental conditions that produce the problems in the first place. Or the pattern of promotion shown in Table 9.1 could be rearranged so that more housing gets built and repaired even if fewer offices and shops result. On any sensible cost/benefit assumptions such a strategy is likely to save the public money spent on repairing people's minds and bodies and in 'soft policing' them away from anti-social behaviour born partly out of boredom or anger at the low quality of their home environment. At present this inherently reasonable suggestion is not being translated into policy. It is difficult to see how it will be so long as the key decision-making processes are moving away from, not towards, the democratic arena.

The final pro-democratic argument is simply that in Britain in the 1980s we have seen the generation of new and renewed built environment carried out largely on the basis of accumulative and competitive motivations, with the regulatory and interventionary hand of the democratic state pro-gressively weakened. As has been argued, there is little to indicate that this strategy has produced either greater efficiency or equity, but plenty of evidence to show that it has produced the reverse. The output of new housing sinks lower and lower. The construction industry is in the doldrums. Low-rent housing is almost unobtainable in many areas. The market in existing housing is depressed in terms both of prices and of volume of transactions. The mobility of the labour force is seriously impeded by the sluggish behaviour of this market. The rents of existing shops and offices have fallen to new lows, and an enormous square footage stands vacant. The low rents achievable by new developments coming on to the market confound the financial calculations of the schemes' promoters – calculations already rendered more bleak by the periodic incidence of high interest rates required by years of myopic macro-economic policy. Few people pretend that it is anything other than a complete mess. It is at least a reasonable supposition that there is some connection between the rapid removal of democratic accountability and the chaotic effects observed. How might this be redressed?

TOWARDS A MORE GENUINE SHARING
OF POWER

It has been argued that the inequality between areas such as those in Chapter 2 of the book has come about not by chance but as a logical outcome of neo-liberal beliefs and policies. It is equally clear that publicly financed and democratically managed services are likely to produce more equitable outcomes than are those produced and sold by the private sector. Thus as publicly organised services are progressively eroded inequalities are likely to grow. There was some evidence of this in Chapter 2, where it was shown that the indicators that exhibited least inequality between the three areas, for example council rents and pupil:teacher ratios, were those measuring the output of services delivered by well-established and democratically accountable public systems of provision. Other key indicators – the rate of unemployment, house-price differentials and the percentage of people on benefits – varied from area to area by factors of three or four. These reflect area-based inequalities in levels of prosperity that flow from investment and disinvestment decisions made in the NDA or capital-accumulative sector of Figure 3.1. This is not of course to argue for a command economy where there is no private sector and all resource decisions are centrally made. This has been tried elsewhere in Europe and has failed. But the data may indicate that degrees of inequality may be lessened by shifting the balance of power over decisions concerning, for example, housing and public-sector-generated employment back in favour of more democratic procedures.

The loss of local democracy has occurred in three main ways: the shifting of more power over local spending programmes to central government, the privatisation of some services, and the shifting of responsibility for others to quangos or some other form of non-elective body. The era of the powerful and autonomous local authority began to fade when the costs of post-war reconstruction meant the need for increasing proportions of central-government funding. Then for two decades the longer-term realities were partly obscured in the optimistic 'never had it so good' period that ended fairly abruptly for Britain in the period leading up to and including the oil crisis of the early 1970s. From the mid-1970s onwards 'the party has been over' for the Town Halls – to quote a pre-Thatcher Labour Secretary of State. In terms of funding and legally defined capabilities it has been all downhill since (see Loughlin, Gelfand and Young, 1985). By the late 1980s, only about 25 per cent of the funds spent by local government were being raised locally (Lansley, Goss and Wolmar, 1989, 184). At a time when the Public Sector Borrowing Requirement is spiralling upwards alarmingly, there is perceived to be even more need to clamp down on local authorities, even Tory ones, who show signs of wishing to provide services in excess of the tight centrally imposed constraints.

One can see the logic for all this – but only if the guiding imperatives are

Plate 11.1 Houses built under the Seaview Self-Build Housing and Training Scheme in Brighton – an example of small-scale and flexible response to housing need. The scheme was set up in 1990 by the Housing Department under a Department of the Environment £400,000 Supplementary Credit Approval. The purpose was to enable homeless 16- and 17-year-olds to build houses for their own use. The development includes sixteen timber framed units built in pairs. The construction technique, designed by Kenneth Claxton, uses standard sized components and requires no 'wet' building skills. The units, which cost about £14,600 to build, are managed by the town's Housing Department. The self-builders all took a two-year course leading to a National Vocational Qualification and developed vocational and personal skills. Photograph by Peter Ambrose.

the necessity to balance budgets year by year, to cut public expenditure to the bone and to show especial evidence of 'good housekeeping' as elections loom. But these imperatives stem from short-term thinking and the definition of costs and benefits in inappropriately narrow terms. There is as yet no serious attempt by policy-makers or advisers to assess the likely costs and benefits of a more medium- or long-term strategy towards the built environment. The understanding that infrastructural investment helps an economy has not yet been extended to encompass the notion that good housing and environmental conditions also count as infrastructure. They are important, perhaps crucial, elements in improving the performance of the economy in the medium and long term because they facilitate the development of the total pool of talent in the society. They also affect future health, welfare and policing expenditure, probably on a massive scale. These arguments, so clearly understood by Addison in 1922 (see Chapter 6) have been lost sight of by today's less visionary politicians.

Similarly, sensible land-allocation and development policies, carefully

216

integrated with public transport policies, will gradually reduce energy, environmental and congestion costs by reducing the aggregate amount of movement necessary for an urban area to function effectively. The Dutch planning system recognises this truth. Many British planners can see the point too, but their capacity to fashion the environment has been much eroded under neo-liberal governments. The longer view in these respects cannot emerge under the *laissez-faire*, free-for-all and speculative approach to the built environment of which the London Docklands case is so graphic an example. It is crucially important, economically as well as socially, that we move back towards a more balanced approach to the processes set out in Figure 3.1.

ALL A QUESTION OF SCALE AND BALANCE?

If there seems at present no obvious way that local people's influence over events can be reliably restored via the ballot box in local elections, can users of the built environment participate in other ways in key decisions affecting their lives? Maybe one way forward lies somewhere in the third mechanism noted above – in the application of individual energies and initiative via the further development of voluntary-sector organisations. There are dangers here of the paternalism, or maternalism, of an Octavia Hill (Chapter 7), and no doubt some large voluntary organisations can be at least as insensitive to user needs as some local authorities. But maybe the secret lies in the 'large' – in the scale of operation. Perhaps in the end it doesn't matter whether a providing organisation is in the 'public' or the 'private' sector. The model presented as Figure 3.1, helpful as it may or may not be in unpacking the process producing built environment, has focused on categorising agencies on the basis of ownership and formal democratic accountability. Maybe this is mistaken and it might be more productive to focus on scale, flexibility and willingness to respond to user needs.

In this context it is vital to ensure that the voluntary housing organisations which are increasingly responsible for providing 'social housing' (see Chapter 7) do in fact remain sensitive to social as well as economic considerations. The Thatcher message was that everything should be reduced to balance-sheet values and that narrowly defined 'value for money' for government expenditure was all that need be aimed for. This has placed strong pressures on the voluntary housing movement both to reduce the quality of provision and to increase the scale of developments so as to benefit from short-term 'scale economies'. Subsequently various studies (for example Page, 1993) have drawn attention to the dangers of producing latter-day versions of some large local authority estates where people suffering from problems related to poverty were clustered together without the necessary community facilities and support being made available to help them deal with the problems and maintain and improve the physical and social fabric around them.

To avoid the serious effects that have been seen to emerge from these conditions elsewhere it is important that those who run housing associations, and the Housing Corporation itself, should make clear the opposition they feel to over-large, socially homogeneous and poor-quality estate development. Otherwise the voluntary housing movement, for all its high ideals and long experience, may be corrupted into becoming yet another instrument for marginalising the poor. Interestingly, it is the commercially driven lenders who are most opposed to the drift of policy here. The government seems mainly concerned with the short-term aim of maximising the amount of development per unit of grant provided while remaining oblivious to the longer-term costs that will inevitably ensue. The private-sector lenders have a longer-term view. They are aware that the future value of the assets they are financing depends upon the quality of development and the attraction of socially mixed and mutually supportive resident populations. This is a small-scale and specific example of the general argument made elsewhere in the book that a proper level of investment in the built environment, and especially in housing, is an investment in the quality of the whole society and that it is a correct strategy both morally and in terms of cost-effectiveness.

Of the organisations discussed in this book those that appear most responsive to user needs, at least so far as housing is concerned, seem to be Brighton Housing Trust (discussed in Chapter 7), the Swedish *kollektivhuset* in Örebro and the Danish co-operatives (both discussed in Chapter 10). They are all fairly small organisations and in the latter two at least there is an obvious opportunity for residents to influence not only management decisions affecting their conditions of occupancy but also resource issues such as selling-prices and rents. Within the democratic sector too the Swedish arrangements seem to be responsive to local preferences, since changes in the political control of communes leads to changes in development policy in a way that is surprising to British observers.

Another feature of many European development systems that may point a way ahead for Britain is the extent to which self-help housing arrangements are developed (Chapter 7). This is a means by which households can play a large part in fashioning their immediate living environment either individually or collectively. Again the operation is essentially small-scale at the point of delivery, but its viability depends on the supportive environment provided by democratic local authorities. These suggestions about ways to regain some user power all relate to the residential environment. There seems to be no obvious way that the major resource decisions influencing the pattern of economic activity can become more participative. Multi-nationals have become a law unto themselves.

This last part of the book is about lessons. What broad conclusions can we draw from reflecting on the various areas and development situations touched on in the book – on Tower Hamlets, Knowsley, Wokingham, Docklands, Downham and Bellingham, Brighton, Toulouse, Stockholm,

Örebro, Bulgaria, Denmark, the City of London and the rest? What can we learn from ministerial pronouncements and attitudes which have ranged from the pragmatism of Lloyd George and the vision of Addison and Bevan to the myopic and near-totalitarian stance of recent neo-liberals? Perhaps the main lesson is that control over the shaping of the built environment is too important economically, socially and ecologically to be left so comprehensively to market-driven organisations. These are often short-term in their thinking, speculative in approach, sometimes massively wasteful of other people's money, and they have a tendency to self-destruct. More especially is this so when government, as has happened, abdicates from its role as market regulator and guardian of the longer-term public interest. In these conditions the *laissez-faire* approach has frequently not worked even in its own terms and it has certainly produced greater inequalities in living conditions. We must surely seek a better balance of private, public and voluntary influences in the evolution of the built environment. In other words we have come full circle back to the Introductory Note.

ANNOTATED FURTHER READING

2 THREE URBAN ENVIRONMENTS COMPARED

As has been evident, much of the data in this chapter has been derived from the *County Reports* of the Census which is taken every ten years in the year ending in –1. Because of the wide range of data they contain, these reports take up to two or more years to publish and the data within them subsequently becomes dated. The *County Monitors*, also produced by the Office of Population Censuses and Surveys, contain a narrower range of county data in an easily accessible form. When starting a search for officially compiled data it is often best to go first to *Social Trends*, or for regional data to *Regional Trends*. Both are produced annually by the Central Statistical Office. These publications not only include chapters presenting data on the most important areas of social and economic analysis but also direct the researcher to the other official sources drawn on. Important social and economic research is carried out continually by a number of research centres and foundations, including the Policy Studies Institute, the Nuffield Foundation and the Joseph Rowntree Foundation. These should be contacted for publication lists and other free or cheap material – for example, the *Findings* series of the latter foundation. Shelter carries out continual research on housing issues and both the Nationwide Anglia and the Halifax Building Societies provide free material on house-price trends. The *Employment Gazette*, published monthly, gives data on employment and unemployment, and the Low Pay Unit should be contacted for their list of publications concerning research on incomes.

3 THE SYSTEM GENERATING NEW BUILT ENVIRONMENT

So far as the author is aware, no model similar to that presented in this chapter exists in the urban-development literature, although various models have been critically reviewed by Gore and Nicholson in *Environment and Planning A*, 23 (1991). If pressed to choose three really seminal pieces of work on which to base the development of an understanding of 'urban process' the choice would be Lamarche's essay in *Urban Sociology: Critical Essays*, edited by Pickvance; chapter 12 of Harvey's

The Limits to Capital (for the more theoretically inclined); and as much as possible of Mumford's great historical sweep *The City in History*.

4 PROFIT-SEEKING DEVELOPMENT – AS INVESTMENT

The most recent and authoritative book on investment property development is the third edition (1991) of *Property Development* by Cadman and Austin-Crowe (this edition is edited by Topping and Avis). Also very useful is the fourth edition of Balchin *et al.*, *Urban Land Economics and Public Policy*. Both these books are part-authored by property practitioners, and they go into considerable detail about all aspects of the development process. *Industrial and Business Space Development* by Morley *et al.* concentrates primarily, as the title implies, on industrial development. *The Dynamics of Urban Property Development* by Rose, one of Britain's most experienced developers, is a readable account that seeks to cover a historical period from the Industrial Revolution to the present. Finally, no one seeking to understand the early post-war era of development should miss Marriott's *The Property Boom* – a racy account of the more buccaneering era before the days of complex computer-based development appraisals. For information on trends in rents and yields in the various development categories, one can approach one of the main consultancy/advisory agents such as Hillier Parker or Debenham Tewson Research. If the thirster after knowledge looks poor enough and academic enough (much the same thing really), she or he may get some information for nothing.

5 PROFIT-SEEKING DEVELOPMENT – FOR SALE

Merrett and Gray's book *Owner Occupation in Britain*, dating from 1982, is a rigorous survey of the history of home ownership up to the late 1970s. Much has happened since then and it is examined, from rather differing perspectives, by Saunders in *A Nation of Home Owners* and by Forrest, Murie and Williams in *Home Ownership: Differentiation and Fragmentation*, both published in 1990. Various knowledgeable works by Ball, including *Housing Policy and Economic Power*, give excellent accounts of the structure and workings of the housebuilding industry. The second edition of Burnett's *A Social History of Housing 1815–1985*, and Jackson's *Semi-Detached London* both range broadly over housing history but give especially well-illustrated accounts of the growth of owner occupation between the wars and subsequently. There is an extensive literature on the building society movement, including Boddy's *The Building Societies* and Boleat's *The Building Society Industry*. The Council of Mortgage Lenders produces much authoritative information about the private-housing market, and Credit Lyonnais Laing publishes surveys on the housebuilding industry from time to time. Finally the Shelter house journal *Roof*, published six times per year, often includes well-informed short articles on the owner-occupancy sector.

6 NON-PROFIT SEEKING DEVELOPMENT – STATUTORY

Perhaps the best and most thorough account of council housing in Britain, up to the late 1970s, is Merrett's *State Housing in Britain*. This study has both historical depth and analytical rigour. An excellent brief review of the early history of subsidy arrangements can be found in the National CDP publication *Whatever Happened to Council Housing?* Malpass and Murie's *Housing Policy and Practice* is comprehensive and authoritative, and the second edition takes the story up to the mid-1980s. Other works by these two authors, including *Reshaping Housing Policy: Subsidies, Rents and Residualisation* by Malpass, can be relied upon to be careful and balanced in their treatment. *Implementing Housing Policy*, edited by Malpass and Means, is a recent (1993) collection of very useful contributions, and the 1991 book *Housing Finance in the U.K.*, edited by Gibb and Munro, contains much of value. Other works that should be consulted include Swenarton's excellent historical work *Homes Fit for Heroes*, Holmans's *Housing Policy in Britain*, and *Public Housing: Current Trends and Future Developments* edited by Clapham and English.

7 NON-PROFIT-SEEKING DEVELOPMENT – VOLUNTARY

The literature on the voluntary housing sector is not easily summarised. Cope's 1990 book *Housing Associations: Policy and Practice* is a useful guide to those aspects summarised by its title. Best's chapter in *A New Century of Social Housing*, edited by Lowe and Hughes, gives a good historical summary. Black and Hamnett's 1985 paper in *Policy and Politics* is an excellent discussion of the voluntary sector's changing political role. The Housing Corporation has produced many useful publications including *Into the Nineties* and *Housing Associations in 1992*. It also publishes an *Annual Report*, which contains information on activities and finances. Similarly the National Federation of Housing Associations produces frequent research reports on the activities of the voluntary sector. Numbers 14–17 on various effects of the Housing Act, 1988 are especially important in discussing recent changes in the sector's role. Jones's 1985 report *The Jubilee Album*, also published by the National Federation, gives a vivid and well-illustrated account both of the history of the sector and of its present diversity. Finally most of the standard texts on housing listed in the Bibliography, such as those by Malpass and Murie, Malpass and Means, Gibb and Munro, and M. Smith contain useful chapters on the voluntary sector.

8 THE DOMINANT 'NEO-LIBERAL' IDEOLOGIES OF THE 1980s/1990s

The production of literature on Mrs Thatcher and 'Thatcherism' has been one of Britain's few growth industries during the 1980s. A good starting-point is Green's *The New Right: The Counter Revolution in Political, Economic and Social Thought*, whose title says it all, or Bosanquet's earlier *After the New Right*. Both are now a little dated. These books, and a number of others, summarise some aspects of the thought of key neo-liberal authors, but in addition it is important to read work by Hayek, starting perhaps with *The Road to Serfdom*, first published in 1944, and by Friedman, perhaps *Capitalism and Freedom*. Chapter 8 of Dearlove and Saunders's *Introduction to British Politics* provides a valuable guide to the thinking of these, and other, neo-liberals. Johnson's *The Economy under Mrs Thatcher 1979–1990* and Wilson's *A Very British Miracle: The Failure of Thatcherism* are both well evidenced and highly critical accounts of Britain's economic performance in the Thatcher years. Gamble's *The Free Economy and the Strong State* is another invaluable review and critique of the Thatcher governments until 1987, and Andrews and Jacobs's *Punishing the Poor* discusses just that. Finally there are two books by respected journalists that should be consulted: Young's *One of Us: The Final Edition*, which traces Mrs Thatcher's remarkable rise to power, her period of power and her eventual fall, and Riddell's *The Thatcher Era and its Legacy*.

9 THE IMPACT OF 'NEO-LIBERAL' POLICIES ON THE BUILT ENVIRONMENT

The first part of this chapter was seeking to develop an argument about the relationship between changes in dominant politics and the effects on patterns of production of new built environment. So far this analysis is little developed in the literature, but some of the chapters in *Policy and Change in Thatcher's Britain*, edited by Cloke, throw some light on the issue. Those who wish to pursue it should be aware of the main sources of relevant statistics, including *Housing and Construction Statistics* and various statistics on housing produced by the Chartered Institute of Public Finance Accountants (CIPFA). On the question of how local authority rents are set the standard works on housing finance, such as Aughton's *Housing Finance: A Basic Guide*, Hills' *Unravelling Housing Finance* and Gibb and Munro's *Housing Finance in the UK: An Introduction*, are all useful. To keep up to date with rent levels in the public sector, use can be made of the statistical series mentioned above or of the data on rents that appear from time to time in Shelter's publication *Roof*. On the London Docklands redevelopment there is a rich and growing literature: Brownill's *Developing London's Docklands: Another Great Planning Disaster?* is excellent, and the publications by the Docklands Consultative Committee, including *All That Glitters is Not Gold*, are well informed and constitute collectively a strong critique. *London Docklands: the Challenge of Development*, edited by Ogden, contains an extremely useful collection of short papers on various aspects of the redevelopment. Finally Widgery's *Some Lives* gives a firsthand account of the difficulties of life in the inner city.

10 HOW IT WORKS ELSEWHERE

The comparative material discussed in this chapter has so far been published more extensively in academic journals than in book form. One exception to this is the 1985 book *Housing, States and Localities* edited by Dickens *et al.* The 1992 monograph in *Progress in Planning* by Barlow and Duncan summarises fully the three-nation comparative-research project discussed, and further aspects of this study are discussed in Duncan and Barlow's 1991 report for the Swedish Council for Building Research. Clapham and Kintrea's 1987 paper in *Housing Studies* is a useful comparative paper on housing co-operatives. Elander and Stromberg's chapter in *Policy, Organization, Tenure: A Comparative History of Small Welfare States*, edited by Lundquist, deals with post-1991 changes in Swedish politics and their effects on land-development policies. All these works are listed in the Bibliography. The housing situation in Bulgaria as at an early stage of the reform process is discussed in *The Reform of Housing in Eastern Europe and the Soviet Union*, edited by Turner *et al.* This is useful as a 'snapshot' but may soon become outdated. Events are moving so fast in the reforming economies that it may be a few years before a systematic and balanced account of the process in book form can be recommended.

BIBLIOGRAPHY

Abel-Smith, B. and Townsend, P. (1965) *The Poor and the Poorest*, London, Bell (Occasional Papers in Social Administration, V. 17)

Addison, C. (1922) *The Betrayal of the Slums*, London, Herbert Jenkins

Advisory Group on Commercial Property Development (1975) *Commercial Property Development* (the 'Pilcher Report'), London, HMSO

Alexander, C. (1979) *The Timeless Way of Building*, Oxford, Oxford University Press

Allen, J. and Hamnett, C. (1991) *Housing and Labour Markets: Building the Connections*, London, Unwin Hyman

Ambrose, P. (1976) *The Land Market and the Housing System*, Brighton, University of Sussex (Urban and Regional Studies Working Paper 3)

Ambrose, P. (1977) *The Determinants of Urban Land Use Change*, Unit 26 in Open University Course D204, Fundamentals of Human Geography

Ambrose, P. (1986) *Whatever Happened to Planning?*, London, Methuen

Ambrose, P. (1991) 'The housing provision chain as a comparative analytical framework', *Scandinavian Housing and Planning Research*, 8.2: 91–104

Ambrose, P. (1992) 'The performance of national housing systems – a three-nation comparison', *Housing Studies*, 7.3: 163–76

Ambrose, P. and Barlow, J. (1986) *Housing Provision and Housebuilding in Western Europe: Increasing Expenditure, Declining Output?*, Brighton, University of Sussex (Urban and Regional Studies Working Paper 50)

Ambrose, P. and Colenutt, B. (1975) *The Property Machine*, Harmondsworth, Penguin

Andrews, K. and Jacobs, J. (1990) *Punishing the Poor: Poverty under Thatcher*, London, Macmillan

Ash, J. (1980) 'The rise and fall of high-rise housing in England', in C. Ungerson and V. Karn, eds, *The Consumer Experience of Housing*, London, Gower

Ashworth, W. (1954) *The Genesis of Modern British Town Planning*, London, Routledge & Kegan Paul

Atkinson, A. (1975) *The Economics of Inequality*, Oxford, Clarendon Press

Aughton, H. (1990) *Housing Finance: A Basic Guide*, 3rd revised edition, London, Shelter

Back, G. and Hamnett, C. (1985) 'State housing policy formation and the changing role of housing associations', *Policy and Politics*, 13.4: 393–411

Bacon, R. and Eltis, W. (1976) *Britain's Economic Problem: Too Few Producers*, London, Hutchinson

Bailey, R. (1992) 'DIY for the homeless', in C. Grant, ed., *Built to Last?*, London, Shelter

Baker, C. (1976) *Housing Associations*, London, Estates Gazette Ltd

Balchin, P. (1989) *Housing Policy: An Introduction*, London, Routledge

Balchin, P., Kieve, J. and Bull, G. (1988) *Urban Land Economics and Public Policy*, 4th edition, London, Macmillan

Ball, M. (1983) *Housing Policy and Economic Power*, London, Methuen

Ball, M. (1986) *Home Ownership: A Suitable Case for Treatment*, London, Shelter

Ball, M. (1988) *Rebuilding Construction*, London, Routledge

Ball, M. (1990) *Under One Roof*, London, Routledge

Banco di Hipotecario de España (1989) *Evolucion de los Precios de la Vivienda*, Madrid, Banco di Hipotecario de España

Barlow, J. (1993) 'Controlling the land market: some examples from Britain, France and Sweden', *Urban Studies* (forthcoming)

Barlow, J., Ambrose, P. and Duncan, S. (1988) 'Housing provision in high growth regions', *Scandinavian Housing and Planning Research*, 5: 33–8

Barlow, J. and Duncan, S. (1988) 'The uses and abuses of housing tenure', *Housing Studies*, 3.4: 219 31

Barlow, J. and Duncan, S. (1992) 'Markets, states and housing provision: four European growth regions compared', *Progress in Planning*, 38.2: 94–177

Barlow, J. and King, A. (1992) 'The state, the market and competitive strategy: the housebuilding industry in the United Kingdom, France and Sweden', *Environment and Planning A*, 24: 381–400

Barlow, J. and Savage, M. (1991) 'Housing the workers in Mrs Thatcher's high-tech Utopia', in J. Allen and C. Hamnett, eds, *Housing and Labour Markets: Building the Connection*, London, Unwin Hyman

Barnes, P. (1984) *Building Societies: The Myth of Mutuality*, London, Pluto Press

Barrass, R. (1984) 'The office development cycle in London', *Land Development Studies*, 1: 35–50

Barrett, S. and Healey, P. (1985) *Land Policy: Problems and Alternatives*, Aldershot, Gower

Bassett, K. (1976) *Public Housing in Bristol 1918-1939: A Study of National Policies and Local Response*, Bristol, University of Bristol Department of Geography (Bristol Housing Studies no. 1)

Batley, R. (1989) 'London Docklands: an analysis of power relations between UDCs and local government', *Public Administration*, 67: 167–87

Bell, E. (1942) *Octavia Hill*, London, Constable

Bellman, H. (1928) *The Silent Revolution*, London, Methuen

Best, R. (1991) 'Housing associations: 1890–1990', in S. Lowe and D. Hughes, eds, *A New Century of Social Housing*, Leicester, Leicester University Press

Beveridge, W. (1942) *Social Insurance and Allied Services*, London, HMSO

Birch, J. (1993) 'Less value, less money', *Roof*, March/April, 26–8

Birchall, J., ed. (1992) *Housing Policy in the 1990s*, London, Routledge

Black, G. and Hamnett, C. (1985) 'State policy formation and the changing role of housing associations in Britain', *Policy and Politics* 13.4: 393–411

Boddy, M. (1980) *The Building Societies*, London, Macmillan

Boddy, M. (1989) 'Financial deregulation and UK housing finance: government building society relations and the Building Societies Act 1986', *Housing Studies*, 4.2: 92–104

Boleat, M. (1982) *The Building Society Industry*, London, Allen & Unwin

Boleat, M. (1985) *National Housing Finance Systems*, London, Croom Helm

Booth, W. (1890) *Darkest England, or The Way Out*, London, Salvation Army

Bosanquet, N. (1983) *After the New Right*, London, Heinemann

Bournville Village Trust (1941) *When We Build Again*, London, Allen & Unwin

Bowley, M. (1945) *Housing and the State*, London, Allen & Unwin

Bramley, G. (1991) *Public Sector Housing Rents and Subsidies*, University of Bristol, School for Advanced Urban Studies (Working Paper 92)

Bramley, G. (1993) 'The enabling role for local housing authorities: a preliminary evaluation', in P. Malpass and R. Means, eds, *Implementing Housing Policy*, Buckingham, Open University Press

Branson, N. (1979) *Popularism 1919–26; George Lansbury and the Councillors' Revolt*, London, Lawrence & Wishart

Braunbehrens, V. (1991) *Mozart in Vienna*, Oxford, Oxford University Press

Briggs, A. (1961) *Social Thought and Social Actions: A Study of the Work of Seebohm Rowntree 1871–1954*, London, Longman

Brimacombe, M. (1991) 'Homes fit for heroes?', *Roof*, January/February, 22–5

Brindley, T., Rydin, Y. and Stoker, G. (1989) *Remaking Planning: The Politics of Urban Change in the Thatcher Years*, London, Unwin Hyman

Brown, G. (1989) *Where There is Greed . . .*, Edinburgh, Mainstream

Brownill, S. (1990) *Developing London's Docklands: Another Great Planning Disaster?*, London, Paul Chapman

Bruce, M. (1968) *The Coming of the Welfare State*, London, Batsford

Bryman, A. *et al.*, eds (1987) *Rethinking the Life Cycle*, London, Macmillan

Budd, L. and Whimster, S. (1992) *Global Finance and Urban Living*, London, Routledge

Budge, I. and McKay, D. (1993) *The Developing British Political System: The 1990s*, London, Routledge

Burnett, J. (1991) *A Social History of Housing 1815–1985*, London, Routledge

Burrows, L. (1989) *The Housing Act 1988*, revised edition, London, Shelter

Burrows, L., Phelps, L. and Walentowicz, P. (1993) *For Whose Benefit? The Housing Benefit Scheme Reviewed*, London, Shelter

Byrne, D. *et al.* (1986) *Housing and Health: The Relationship between Housing Conditions and the Health of Council Tenants*, London, Gower

Byrne, P. and Cadman, D. (1984) *Risk, Uncertainty and Decision-Making in Property Development*, London, E. and F. N. Spon

Cadman, D. (undated) 'Visions of the city', London, Property Market Analysis (unpublished paper)

Cadman, D. and Austin-Crowe, L. (1991) *Property Development*, 3rd edition, London, E. and F. N. Spon

Central Housing Advisory Committee (1939) *The Operation of Housing Associations*, London, HMSO

Central Statistical Office (various) *Financial Statistics*, London, Central Statistical Office

Centre for Urban and Regional Studies (1981) *The Housebuilding Industry and Changes in the Market for Housebuilding Work: A Review of the British Experience by the Housing Monitoring Team*, Birmingham, University of Birmingham (Research Memorandum 87)

Chapman, S., ed. (1971) *The History of Working Class Housing*, Newton Abbott, David & Charles

Church, A. (1988) 'Urban regeneration in London's Docklands: a five-year policy review', *Environment and Planning C: Government and Policy*, 6: 187–208

Clapham, D. (1991) 'Sweet charity', *Roof*, January/February, 38–41

Clapham, D. and English, J., eds (1987) *Public Housing: Current Trends and Future Developments*, London, Croom Helm

Clapham, D., Kemp, P. and Smith, S. (1990) *Housing and Social Policy*, London Macmillan

Clapham, D. and Kintrea, K. (1987) 'Importing housing policy: housing co-operatives in Britain and Scandinavia', *Housing Studies*, 2.3: 157–69

Cloke, P., ed. (1992) *Policy and Change in Thatcher's Britain*, Oxford, Pergamon Press

Coakley, J. (1992) 'London as an international financial centre', in L. Budd and S. Whimster, eds, *Global Finance and Urban Living*, London, Routledge

Cochrane, A. (1993) *Whatever Happened to Local Government?*, Buckingham, Open University Press

Collison, P. (1963) *The Cutteslowe Walls: A Study in Social Class*, London, Faber

Conservative Central Office (1979) *The Sale of Council Homes – Model Scheme, Guidance Notes*, London, Conservative Central Office

Cooke, C. (1957) *The Life of Richard Stafford Cripps*, London, Hodder & Stoughton

Cooke, P. (1989) *Localities: The Changing Face of Urban Britain*, London, Unwin Hyman

Cooney, C. (1974) 'High flats in local authority housing in England and Wales since 1956', in A. Sutcliffe, ed., *Multi-Storey Living*, London, Croom Helm

Cope, H. (1990) *Housing Associations: Policy and Practice*, London, Macmillan

Counsell, G. (1990) 'Clinging to the wreckage', *Roof*, May/June, 26–7

Counter Information Services (1973) *The Recurrent Crisis of London*, London, Counter Information Services

Coupland, A. (1992) 'Docklands: dream or disaster?' in A. Thornley, ed., *The Crisis of London*, London, Routledge

Coxall, W. (1990) *Parties and Pressure Groups*, 2nd edition, London, Longman

Craig, P. (1986) 'The house that Jerry built? Building societies, the state and the politics of owner-occupation', *Housing Studies*, 1.1: 87–108

Credit Lyonnais Laing (1991) *Private Housebuilding*, London, Credit Lyonnais Laing

Credit Lyonnais Laing (1993) *Construction and Housebuilding Sector*, London, Credit Lyonnais Laing

CSW (1993) *The Property Week*, 29 July/5 August, 32–4

Cullingworth, J. (1979) *Essays on Housing Policy*, London, Allen & Unwin

Cullingworth, J. (1988) *Town and Country Planning in Britain*, 10th edition, London, Unwin Hyman

Damer. S. (1992) 'Striking out on Red Clyde', in C. Grant, ed., *Built to Last?*, London, Shelter

Danermark, B. and Vinterhennei, S. (1991) 'Housing provision in European growth regions: consumption in the Stockholm–Arlanda region', *Swedish Council for Building Research*, D14

Darley, G. (1990) *Octavia Hill: A Life*, London, Constable

Daunton, M., ed. (1984) *Councillors and Tenants: Local Authority Housing in English Cities, 1919–1939*, Leicester, Leicester University Press

Daunton, M. (1987) *A Property Owning Democracy? Housing in Britain*, London, Faber & Faber

Dawson, A. (1983) *Shopping Centre Development*, London, Longman

Deakin, N. (1987) *The Politics of Welfare*, London, Methuen

Dearlove, J. and Saunders, P. (1991) *Introduction to British Politics*, 2nd edition, Oxford, Polity Press

Debenham Jean Thouard Zadelhoff (1992) *European Commercial Property Markets: An Overview*, London, Debenham Jean Thouard Zadelhoff

Debenham Tewson Research (1990a) *Business Parks – Out of Town or Out of Touch?*, London, Debenham Tewson Research

Debenham Tewson Research (1990b) *Retailing into the 90s: The Multiples' Choice* London, Debenham Tewson Research

Debenham Tewson Research (1990c) *Offices Gloom and Boom*, London, Debenham Tewson Research

Debenham Tewson Research (1992a) *London through the 1990s: Complacency and Competition*, London, Debenham Tewson Research

Debenham Tewson Research (1992b) *Office Demand: Choice and Compromise*, London, Debenham Tewson Research

Debenham Tewson Research (1992c) *Central London Pipeline Rescheduling – Too Much Too Late?*, London, Debenham Tewson Research

Debenham Tewson Research (1993a) *Distribution 2000*, London, Debenham Tewson Research

Debenham Tewson Research, (1993b) *Overseas Investment in UK Commercial Property 1992*, London, Debenham Tewson Research

Department of Social Security (1992) *Households Below Average Incomes 1979–1988/89*, London, HMSO

Dickens, P. (1988) *One Nation? Social Change and the Politics of Locality*, London, Pluto Press

Dickens, P. *et al.* (1985) *Housing, States and Localities*, London, Methuen

Direct Labour Collective (1978) *Building with Direct Labour*, London, Direct Labour Collective

Docklands Consultative Committee (1990) *The Docklands Experiment. A Critical Review of Eight Years of the London Docklands Development Corporation*, London, Docklands Consultative Committee

Docklands Consultative Committee (1992) *All That Glitters is Not Gold*, London, Docklands Consultative Committee

Doling, J. and Stafford, B. (1989) *Home Ownership: The Diversity of Experience*, Aldershot, Gower

Doling, J., Karn, V. and Stafford, B. (1986) 'The impact of unemployment on home ownership', *Housing Studies*, 1.1: 49–59

Donnison, D. and Maclennan, D. (1985) 'What should we do about housing?', *New Society*, 11 (April), 43–6

Duncan, S. (1986) 'Housebuilding, profits and social efficiency in Sweden and Britain', *Housing Studies*, 1.1: 11–33

Duncan, S. (1990) 'Do house prices rise that much? A dissenting view', *Housing Studies*, 5: 195–208

Duncan, S. and Barlow, J. (1991) 'Marketisation or regulation in housing production? Sweden and the Stockholm–Arlanda growth region in European perspective', *Swedish Council for Building Research*, D13

Duncan, S. and Rowe, A. (1992) *Self-Help Housing: The First World's Hidden Housing Arm*, University of Sussex, Centre for Urban and Regional Research (Working Paper 85)

Dunleavy, P. (1981) *The Politics of Mass Housing in Britain 1945–1975*, Oxford, Clarendon Press

Dunleavy, P. (1986) 'The growth of sectoral cleavages and the stabilization of state expenditure', *Environment and Planning D: Society and Space*, 4: 129–44

Dwelly, T. (1992) 'No sign of movement', *Roof*, November/December, 21–5

Dyos, H. (1961) *Victorian Suburb*, Leicester, Leicester University Press

Edel, M. (1982) 'Home ownership and working class unity', *International Journal of Urban and Regional Research*, 6: 205–22

Edmonds, D. (1992) 'How to score a classic own goal', *Roof*, November/December, 15

Elander, I. and Gustafsson, M. (1993) 'The re-emergence of local self-government in

Central Europe', *European Journal of Political Research*, 23: 295–322

Elander, I. and Montin, S. (1990) 'Decentralisation and control: central-local government relations in Sweden', *Policy and Politics*, 18.3: 165–80

Elander, I. and Stromberg, T. (1992) 'Whatever happened to social democracy and planning? The case of local land and housing policy in Sweden', in L. Lundquist, ed., *Policy, Organization, Tenure: A Comparative History of Small Welfare States*, Oslo/Stockholm, Scandinavian University Press

Ellmers, C. and Werner, A. (1991) *Dockland Life: A Pictorial History of London's Docks, 1860–1970*, Edinburgh, Mainstream Publishing

Elsas, M. (1942) *Housing before and after the War*, London, King & Staples

Essex County Council (1973) *Essex Design Guide for Residential Areas*, Essex County Council

European Capital Company Ltd (1991) *Housing Associations: Improved Access to the Capital Markets*, York, Joseph Rowntree Foundation

Feiling, K. (1946) *The Life of Neville Chamberlain*, London, Macmillan

Fielder, S. and Imrie, R. (1986) 'Low cost home ownership: the extension of owner occupation', *Area*, 18: 265–73

Fisher, M. and Owen, U., eds (1991) *Whose Cities?*, Harmondsworth, Penguin

Fisher, M and Worpole, K. (1988) *City Centres, City Cultures*, Manchester, Centre for Local Economic Studies

Forbes, E., ed. (1973) *Thayer's Life of Beethoven*, Princeton, New Jersey, Princeton University Press

Ford, J. (1988) *The Indebted Society: Credit and Default in the 1980s*, London, Routledge

Ford, J. (1992) 'The Damocles sword', *Roof*, July/August, 16–17

Forrest, R. (1983) 'The meaning of home ownership', *Society and Space*, 1. 205–16

Forrest, R. and Murie, A. (1987) 'The affluent home owner: labour market position and the shaping of human histories', *The Sociological Review*, 35.2: 370–403

Forrest, R. and Murie, A. (1989) 'Differential accumulation: wealth, inheritance and housing policy', *Policy and Politics*, 17.1. 25–39

Forrest, R. and Murie, A. (1990) *Selling the Welfare State: The Privatisation of Public Housing*, 2nd edition, London, Routledge

Forrest, R., Murie, A. and Williams, P. (1990) *Home Ownership: Differentiation and Fragmentation*, London, Unwin Hyman

Foster, S. (1993) *Missing the Target*, London, Shelter

Fothergill, S., Monk, S. and Perry, M. (1987) *Property and Industrial Development*, London, Hutchinson

Franklin, A. (1986) *Owner Occupation, Privatisation and Ontological Security: A Critical Reformulation*, Bristol, University of Bristol, School for Advanced Urban Studies (Working Paper no. 62)

Friedman, M. (1982) *Capitalism and Freedom*, Chicago, University of Chicago Press

Gamble, A. (1988) *The Free Economy and the Strong State: The Politics of Thatcherism*, London, Macmillan

Gamble, A. (1990) *Britain in Decline*, 3rd edition, London, Macmillan

Garvey, S. (1991) 'Capital accumulation, local democracy and the state', doctoral thesis, Department of Land Economy, University of Cambridge

Gauldie, E. (1974) *Cruel Habitations; A History of Working-Class Housing 1780–1918*, London, Allen & Unwin

George, V. and Miller, S. (1994) *Social Policy towards 2000: Ignoring the Welfare Circle*, London, Routledge

Gibb, K. and Munro, M. (1991) *Housing Finance in the UK: An Introduction*, London, Macmillan

Giorov, M. and Koleva, M. (1990) 'The current status of Bulgarian housing finance', unpublished paper presented at the Conference on 'Housing Finance in Eastern Europe', Budapest, April

Goodchild, R. and Munton, R. (1985) *Development and the Landowner*, London, Allen & Unwin

Gore, T. and Nicholson, D. (1991) 'Models of the land-development process: a critical review', *Environment and Planning, A*, 23: 705–30

Grant, C., ed. (1992) *Built to Last?*, London, Shelter

Gray, J. (1992) *The Moral Foundation of Market Institutions*, London, Institute of Economic Affairs

Green, D. (1987) *The New Right: The Counter Revolution in Political, Economic and Social Thought*, Brighton, Wheatsheaf Books

Griffiths, P. (1975) *Homes Fit for Heroes*, London, Shelter

Hamer, M. (1987) *The Roads Lobby*, London, Friends of the Earth

Hamnett, C. (1987) 'Conservative government housing policy in Britain, 1979–85', in W. van Vliet, ed., *Housing Markets and Policies Under Fiscal Austerity*, Westport, Connecticut, Greenwood

Hamnett, C. (1991) 'Home ownership, housing and wealth distribution in Britain', paper presented to the 'Housing Policy as a Strategy for Change' conference, Oslo, June

Hamnett, C. and Randolph, W. (1988) *Cities, Housing and Profits*, London, Hutchinson

Hardy, D. (1983a) *Making Sense of London Docklands: Processes of Change* (Papers in Geography and Planning no. 9), Enfield, Middlesex Polytechnic

Hardy, D. (1983b) *Making Sense of London Docklands: People and Places* (Papers in Geography and Planning no. 10), Enfield, Middlesex Polytechnic

Harloe, M. and Martens, M. (1990) *New Ideas for Housing: The Experience of Three Countries*, London, Shelter

Harms, H. (1992) 'Self-help housing in developed and Third World countries', in K. Mathey, ed., *Beyond Self-Help Housing*, London, Mansell

Harris, R. (1991) 'Is London Overdeveloped?', paper presented at the British Sociological Association, 18 May

Harrison, J. (1992) *Housing Associations after the 1988 Act*, Bristol, University of Bristol, School for Advanced Urban Studies (Working Paper no. 108)

Harvey, D. (1982) *The Limits to Capital*, Oxford, Blackwell

Harvey, D. (1985) *The Urbanisation of Capital*, Oxford, Blackwell

Harvey, D. (1989) *The Condition of Postmodernity*, Oxford, Blackwell

Hayek, F. (1944) *The Road to Serfdom*, London, Routledge & Kegan Paul

Hayek, F. (1967) *Studies in Philosophy, Politics and Economics*, London, Routledge & Kegan Paul

Hayek, F. (1973) *Law, Legislation and Liberty*, Vol. I, *Rules and Order*, London, Routledge & Kegan Paul

Hayek, F. (1976) *Law, Legislation and Liberty*, Vol. II, *The Mirage of Social Justice*, London, Routledge & Kegan Paul

Healey, P. and Barrett, S. (1990) 'Structure and agency in land development processes: some ideas for research', *Urban Studies*, 27.1: 89–104

Healey, P. and Nabarro, R., eds (1990) *Land and Property Development in a Changing Context*, Aldershot, Gower

Hegedus, J. (1987) 'Reconsidering the roles of the state and market', *International Journal of Urban and Regional Research*, 12: 79–95

Hillier Parker (various) *Specialised Property Summaries*, London, Hillier Parker

Hills, J. (1987) *The Voluntary Sector in Housing: The Role of British Housing*

Associations, London, Suntory Toyota International Centre for Economics and Related Disciplines, London School of Economics (Working Paper no. 20)

Hills, J. (1991) *Unravelling Housing Finance*, Oxford, Oxford University Press

Hills, J. *et al.* (1990) 'Shifting subsidy from bricks and mortar to people: experiences in Britain and West Germany', *Housing Studies*, 5: 147–69

Hinton, C. (1987) *Using Your Home as Capital*, Mitcham, Age Concern

HMSO (annually) *Housing and Construction Statistics*, London

Hoffman, M. (1991) 'A preliminary assessment of the Bulgarian housing sector', Washington, DC, The Urban Institute (unpublished paper)

Hoffman, M. *et al.* (1992) *The Bulgarian Housing Sector: An Assessment*, Washington, DC, The Urban Institute

Holmans, A. (1987) *Housing Policy in Britain*, London, Croom Helm

Housing Corporation (1990) *Into the Nineties,* London, Housing Corporation

Housing Corporation (1992a) *Annual Report 1991/1992*, London, Housing Corporation

Housing Corporation (1992b) *Housing Associations in 1992*, London, Housing Corporation

Housing Finance Corporation, The (1993) *Group Report and Accounts 1993*, London, The Housing Finance Corporation

Hughes, N. (1990) *Development under the 1988 Act*, London, National Federation of Housing Associations

Hyndman, S. (1990) 'Housing dampness and health among British Bengalis in East London', *Social Science and Medicine*, 30.1: 131–41

Ineichen, B. (1993) *Homes and Health*, London, Spon

Inquiry into British Housing (1985) *Report*, London, National Federation of Housing Associations

Inquiry into British Housing (1986) *The Evidence,* London, National Federation of Housing Associations

Inquiry into British Housing (1991) *Second Report*, London, National Federation of Housing Associations

Jackson, A. (1973) *Semi-Detached London*, London, Allen & Unwin

Jessop, B., Bromley, S. and Ling, T. (1988) *Thatcherism: A Tale of Two Nations*, Cambridge, Polity Press

Johnson, C. (1991) *The Economy under Mrs Thatcher 1979–1990*, Harmondsworth, Penguin

Johnston, R. (1987) 'A note on housing tenure and voting in Britain 1983', *Housing Studies*, 2.2: 112–21

Jones, K. (1991) *The Making of Social Policy in Britain 1830–1990*, London, Athlone Press

Jones, P. (1985) *The Jubilee Album*, London, The National Federation of Housing Associations

Jones, Lang, Wootton (various) *50 Centres*, London, Jones, Lang, Wootton

Joseph, Sir K. and Sumption, J. (1978) *Equality*, London, John Murray

Kaldor, N. (1983) *The Economic Consequences of Mrs Thatcher*, London, Duckworth

Kearns, A. and Maclennan, D. (1989) *Public Finance for Housing in Britain*, Glasgow, University of Glasgow Centre for Housing Research (Discussion paper 22)

Kemeny, J. and Thomas, A. (1984) 'Capital leakage from owner-occupied housing', *Policy and Politics*, 12.1: 1–12

Kemp. P., ed. (1988) *The Private Provision of Rented Housing*, Aldershot, Avebury

Kemp, P. (1993) 'Rebuilding the private rented sector?', in P. Malpass and R. Means, eds, *Implementing Housing Policy*, Buckingham, Open University Press

Key, T. *et al.* (1990) 'Prospects for the property industry: an overview', in P. Healey and R. Nabarro, eds, *Land and Property Development in a Changing Context*, Aldershot, Gower

Koleva, M. and Dandolova, I. (1992) 'Housing reforms in Bulgaria: myth or reality?', in B. Turner, J. Hegedus and I. Tosics, eds (1992) *The Reform of Housing in Eastern Europe and the Soviet Union*, London, Routledge

Lansley, S., Goss, S. and Wolmar, C. (1989) *Councils in Conflict: The Rise and Fall of the Municipal Left*, London, Macmillan

Lewis, J. Parry (1965) *Building Cycles and Britain's Growth*, London, Macmillan

London County Council (1928) *Housing: With Particular Reference to Post-War Housing Schemes*, London, London County Council

London Docklands Development Corporation (various) *Annual Reports*, London, London Docklands Development Corporation

London Docklands Development Corporation (1992) *Key Facts and Figures*, London, London Docklands Development Corporation

Loney, M. *et al.*, eds (1991) *The State or the Market*, 2nd edition, London, Sage

Loughlin, M., Gelfand, M. and Young, K. (1985) *Half a Century of Municipal Decline 1935–1985*, London, Allen & Unwin

Macdonald, L. (1983) *Somme*, London, Macmillan

Maidment, R. and Thompson, G. (1993) *Managing the United Kingdom: An Introduction to its Political Economy and Public Policy*, London, Sage

Maizels, J. (1961) *Two to Five in High Flats*, London, Housing Centre Trust

Malpass, P. (1990) *Reshaping Housing Policy: Subsidies, Rents and Residualisation*, London, Routledge

Malpass, P. (1991) 'Cash till eternity', *Roof*, July/August, 34–7

Malpass, P. (1992) 'Investment strategies', in C. Grant, ed., *Built to Last?*, London, Shelter

Malpass, P. and Means, R. eds (1993) *Implementing Housing Policy*, Buckingham, Open University Press

Malpass, P. and Murie, A. (1987) *Housing Policy and Practice*, 2nd edition, London, Macmillan

Malpass, P. and Warburton, M. (1993) 'The new financial regime for local authority housing', in P. Malpass and R. Means, eds, *Implementing Housing Policy*, Buckingham, Open University Press

Mann, J. and Smith, R. (1993) *Who Says There's No Housing Problem?*, 2nd edition, London, Shelter

Marriott, O. (1967) *The Property Boom*, London, Pan

Marshall, T. (1950) *Citizenship and Social Class and other Essays*, Cambridge, Cambridge University Press

Mathey, K. (1992) *Beyond Self-Help Housing*, London, Mansell

Matrix (1984) *Making Space: Women and the Man-Made Environment*, London, Pluto Press

Mayes, D. (1979) *The Property Boom*, London, Martin Robertson

McCulloch, A. (1990) 'A millstone round your neck? Building societies in the 1930s and mortgage default', *Housing Studies*, 5.1: 43–58

McIntosh, A. and Utley, C. (1992) 'Transferring allegiance', *Roof*, July/August, 18–20

Mearns, A. (1883) *The Bitter Cry of Outcast London: An Enquiry into the Condition of the Abject Poor*, London, London Congregational Union

Meegan, R. (1989) 'Paradise postponed: the growth and decline of Merseyside's outer estates', in C. Cooke, ed., *Localities: The Changing Face of Urban Britain*, London, Unwin Hyman

Merrett, S. (1979) *State Housing in Britain*, London, Routledge & Kegan Paul

Merrett, S. (1991) *Quality and Choice in Housing: A Framework for Financial Reform*, London, Institute for Public Policy Research (Economic Study no. 10)

Merrett, S., with Gray, F. (1982) *Owner Occupation in Britain*, London, Routledge & Kegan Paul

Milton, N. (1973) *John MacLean*, London, Pluto Press

Ministry of Housing and Local Government (1962) *Residential Areas: Higher Densities*, London, HMSO

Morley, S. *et al.* (1989) *Industrial and Business Space Development*, London, E. and F. N. Spon

Morton, J. (1991) *Cheaper than Peabody: Local Authority Housing from 1890 to 1919*, York, Joseph Rowntree Foundation

Muellbauer, J. (1990) *The Great British Housing Disaster and Economic Policy*, London, Institute for Public Policy Research (Economic Study no. 5)

Muellbauer, J. (1993) 'Housing's economic hangover', *Roof*, May/June, 16–21

Mullins, D., Niner, P. and Riseborough, M. (1993) 'Large-scale voluntary transfers', in P. Malpass and R. Means, eds, *Implementing Housing Policy*, Buckingham, Open University Press

Mumford, L. (1966) *The City in History*, Harmondsworth, Pelican

Munro, M. (1988) 'Housing wealth and inheritance', *Journal of Social Policy*, 17.4: 417–36

Murie, A. (1989) *Lost Opportunities? Council House Sales and Housing Policy in Britain 1979–89*, Bristol, School for Advanced Urban Studies (Working Paper 80)

National CDP (1976) *Whatever Happened to Council Housing?*, London, Community Development Projects

National Federation of Housing Associations (1992) *Annual Report 1991–1992*, London, National Federation of Housing Associations

National Housing Forum (1989) *Housing Needs in the 1990s*, London, National Housing Forum

National Housing and Town Planning Council (1929) *A Policy for the Slums*, London, National Housing and Town Planning Council

Ogden, P., ed. (1992) *London Docklands: the Challenge of Development*, Cambridge, Cambridge University Press

Oldman, J. (1990) *Who Says There's No Housing Crisis?*, London, Shelter

Oliver, P., Davis, I. and Bentley, I. (1981) *Dunroamin: The Suburban Semi and its Enemies*, London, Barrie & Jenkins

Orbach, L. (1977) *Homes for Heroes*, London, Seeley Service

Ospina, J. (1987) *Housing Ourselves*, London, Shipman

Owens, R. (1992) 'If the hat fits', *Roof*, January/February, 17–19

Page, D. (1993) *Building for New Communities: A Study of New Housing Association Estates*, York, Joseph Rowntree Foundation

Pahl, R. (1984) *Divisions of Labour*, Oxford, Blackwell

Park, R., Burgess, E. and McKenzie, R. (1925) *The City*, Chicago, University of Chicago Press

Parker, R. (1967) *The Rents of Council Houses*, London, Bell & Sons (Occasional Papers in Social Administration no. 22)

Pawley, M. (1978) *Home Ownership*, London, Architectural Press

Pawley, M. (1992) 'A prefab future', in C. Grant, ed., *Built to Last?*, London, Shelter

Pearce, B., Curry, N. and Goodchild, R. (1978) *Land, Planning and the Market*, University of Cambridge, Department of Land Economy (Occasional Paper 9)

Pickvance, C., ed. (1976) *Urban Sociology: Critical Essays*, London, Methuen

Powell, C. (1974) 'Fifty years of progress', *Built Environment*, October, 532–5

Property Market Analyses (1992) *The PMA Forecasting Service*, London, Property Market Analyses

Randolph, B. (1992) *Housing Associations after the Act*, London, National Federation of Housing Associations (Research Report 16)

Randolph, B. (1993) 'The re-privatisation of housing associations', in P. Malpass and R. Means, eds, *Implementing Housing Policy*, Buckingham, Open University Press

Rao, N. (1990) *The Changing Role of Local Housing Authorities*, York, Joseph Rowntree Memorial Trust

Ravetz, A. (1974) *Model Estate*, London, Croom Helm

Reiss, R. (1919) *The Home I Want*, London, Hodder & Stoughton

Richard Ellis (1993) *World Rental Levels*, London, Richard Ellis

Riddell, P. (1991) *The Thatcher Era and its Legacy*, Oxford, Blackwell

Rose, J. (1985) *The Dynamics of Urban Property Development*, London, E. and F. N. Spon

Rosnes, A. (1987) 'Self-built housing: Norwegian experiences', *Scandinavian Housing and Planning Research*, 4: 55–68

Rowe, A. (1989) 'Self-help housing provision: production, consumption, accumulation and policy in Atlantic Canada', *Housing Studies*, 4.2: 75–91

Royal Commission on the State of Towns and Populous Districts (1845) *2nd Report* (British Parliamentary Papers, XVIII)

Royal Institute of British Architects/National Federation of Housing Associations (undated) *Housing Association Design*, London, Royal Institute of British Architects/National Federation of Housing Associations

Rydin, Y. (1986) *Housing Land Policy*, Aldershot, Gower

Rydin, Y. (1993) *The British Planning System: An Introduction*, Basingstoke, Macmillan

Saunders, P. (1990) *A Nation of Home Owners*, London, Unwin Hyman

Saunders, P. and Harris, C. (1988) *Home Ownership and Capital Gains*, Brighton, University of Sussex (Urban and Regional Studies Working Paper no. 64)

Seymour, J. and Girardet, H. (1990) *Far from Paradise*, London, Greenprint

Shelter (1987) *Homes for the Rich Not the Poor*, London, Shelter

Shelter (1990) *Heard or Ignored: Tenant Representation in Housing Associations*, London, Shelter

Short, J., et al. (1986) *Housebuilding, Planning and Community Action*, London, Routledge

Simey, T. et al. (1960) *Charles Booth, Social Scientist*, London, Oxford University Press

Simon, E. (1933) *The Anti-Slum Campaign*, London, Longmans Green

Smith, A. (1776) *The Wealth of Nations*, Everyman Edition, London, Dent, 1977

Smith, M. (1989) *Guide to Housing*, 3rd revised edition, London, The Housing Centre Trust

Smyth, H. (1985) *Property Companies and the Construction Industry in Britain*, Cambridge, Cambridge University Press

Spicker, P. (1992) 'Victorian values', in C. Grant, ed., *Built to Last?*, London, Shelter

Stevens, S. (1993) *Rent Setting after the Act*, London, National Federation of Housing Associations (Research Report 17)

Stoker, G. (1988) *The Politics of Local Government*, London, Macmillan

Stone, P. (1959) 'The economics of housing and urban development', *Journal of the Royal Statistical Society*, Series A, 122.4.

Sullivan, O. (1986) 'Housing movements of the divorced and separated', *Housing Studies*, 1.1: 34–48

BIBLIOGRAPHY

Sutcliffe, A., ed. (1974) *Multi-Storey Living*, London, Croom Helm

Swenarton, M. (1981) *Homes Fit for Heroes*, London, Heinemann

Swenarton, M. and Taylor, S. (1985) 'The scale and nature of the growth of owner occupation in Britain between the wars', *Economic History Review*, 38.3: 373–92

Szelenyi, I. (1989) 'Housing policy in the emergent socialist mixed economy of Eastern Europe', *Housing Studies*, 4: 167–76

Thornley, A. (1991) *Urban Planning under Thatcherism: The Challenge of the Market*, London, Routledge

Thornley, A., ed. (1992) *The Crisis of London*, London, Routledge

Thorns, D. (1981) 'Owner-occupation: its significance for wealth transfer and class formation', *Sociological Review*, 29: 705–28

Todd, M. (1992) 'Building blues', *Roof*, September/October, 20–2

Took, L. and Ford, J. (1987) 'The impact of mortgage arrears on the housing careers of home owners', in A. Bryman *et al.* eds, *Rethinking the Life Cycle*, London, Macmillan

Tosics, I. (1987) 'Privatisation in housing policy: the case of the Western countries and that of Hungary', *International Journal of Urban and Regional Research*, 11: 61–78

Travers Morgan and Partners (1973) *Docklands Redevelopment Proposals for East London*, London, Travers Morgan and Partners

Treanor, D. (1990) *Housing Association Rents*, London, National Federation of Housing Associations

Tucker, E. (1993) 'Fall in unemployment fails to clarify trend', *Financial Times*, 21 May

Turner, B. *et al.*, eds (1992) *The Reform of Housing in Eastern Europe and the Soviet Union*, London, Routledge

Unicef (1993) *Progress of Nations*, London, UK Committee for UNICEF

Valenca, M. (1992) *The Debate on Equity Withdrawal Versus the Housing-Linked Interest Rate Mechanism*, Brighton, University of Sussex, Centre for Urban and Regional Research (Working Paper 84)

Walentowicz, P. (1991) *Development after the Act*, London, National Federation of Housing Associations (Research Report 14)

Walentowicz, P. (1992) *Design Standards after the Act*, London, National Federation of Housing Associations (Research Report 15)

Walker, K. (1991) 'Tenants' democracy in Denmark', unpublished report prepared for the North British Housing Association

White, J. (1992) 'Business out of charity', in C. Grant, ed., *Built to Last?*, London, Shelter

Widgery, D. (1991) *Some Lives! – A GP's East End*, London, Sinclair-Stevenson

Wilding, P. (1973) 'The Housing and Town Planning Act 1919: a study in the making of social policy', *Journal of Social Policy*, 2.4: 317–34

Williams, G. (1990) 'Development niches and specialist housebuilders', *Housing Studies*, 5.1: 14–23

Willmott, P. and Hutchison, R. (1992) *Urban Trends 1: A Report on Britain's Deprived Areas*, London, Policy Studies Institute

Wilson, E. (1992) *A Very British Miracle: The Failure of Thatcherism*, London, Pluto Press

Wirth, L. (1938) 'Urbanism as a way of life', *American Journal of Sociology*, 44, July, 1–24

Wohl, A. (1977) *The Eternal Slum: Housing and Social Policy in Victorian London*, London, Arnold

Yeo, E. and Thompson, E. (1971) *The Unknown Mayhew*, New York, Random House

Young, H. (1993) *One of Us: The Final Edition*, London, Pan

GENERAL INDEX

INDEX OF LEGISLATION